Design and Crosstalk Analysis in Carbon Nanotube Interconnects

P. Uma Sathyakam · Partha Sharathi Mallick

Design and Crosstalk Analysis in Carbon Nanotube Interconnects

Springer

P. Uma Sathyakam
School of Electrical Engineering
Vellore Institute of Technology
Vellore, India

Partha Sharathi Mallick
School of Electrical Engineering
Vellore Institute of Technology
Vellore, India

ISBN 978-981-15-8887-7 ISBN 978-981-15-8888-4 (eBook)
https://doi.org/10.1007/978-981-15-8888-4

This Springer imprint is published by the registered company Springer Nature Singapore Pte Ltd.
The registered company address is: 152 Beach Road, #21-01/04 Gateway East, Singapore 189721, Singapore

Preface

Scaling down of transistors in integrated circuits is inevitable to improve the speed and reduce power consumption. While scaling down of transistors proved to be beneficial, interconnect scaling in ICs led to many reliability problems. Slowly, the performance-defining factor of an IC has clearly moved from transistors to interconnects over the last decade. Currently, copper (Cu) is used as interconnect material in ICs up to 14 nm technology node. However, beyond this node size, Cu can no more be used as it suffers from electromigration, crosstalk, high resistivity, high RC delay and self-heating. So, alternate materials like carbon nanotubes (CNTs) and graphene nanoribbons (GNRs) are proposed as interconnects and widely studied in the past 10 years. Based on modelling and SPICE simulations, both CNTs and GNRs are found to be excellent replacement to Cu interconnects to connect nanoscale devices in ICs. This book examines the crosstalk and thermal analysis of CNT interconnects. New methods are proposed to reduce crosstalk in CNT interconnects. Semiconducting CNTs and small diameter CNTs are proposed around CNT bundle interconnects as they are found to have smaller dielectric constants than traditionally used Si-based dielectrics. Also, emerging ultra low-k dielectric materials such as boron nitride nanotubes, graphene oxide/polyimide composites and metal organic frameworks are proposed for crosstalk reduction in CNT interconnects. Another method of reducing crosstalk is by using triangular CNT (T-CNT) bundles that have least coupling between them, and hence least possible crosstalk. Temperature coefficient of resistance is calculated for CNTs and the corresponding rise in $R_T C$ delay (where R_T is the temperature-dependent resistance) is calculated. Furthermore, CNTs are treated as discrete fractal media at nanoscale. The fractal dimension of armchair CNTs is modelled for the first time in this book. This fractal dimension is useful to understand the heat flow in CNTs using fractional differential equations. Finally, high-speed CNT interconnects are modelled at subthreshold input voltages by incorporating CNT-FETs-based transmission gate driver and load circuits. Temperature dependence on the performance of the proposed interconnects is carried out from 300K to 500K.

Chapter 2 outlines the background and literature survey of CNT interconnect technology. The Luttinger liquid theory and the electron transport models are described, which treat CNTs as nanoscale quantum wires. Various types of CNT interconnects such as SWCNT, MWCNT and mixed CNT bundle interconnects are described. Then, various design metrics of CNT interconnects and signal propagation delay in CNT interconnects are reviewed. Next, the various aspects of crosstalk in CNT interconnects are reviewed. First, the present status and future challenges are outlined. Then, various crosstalk models like aggressor–victim models, miller capacitance effects and crosstalk in three coupled interconnects are discussed. The various sources of crosstalk-induced peak noise are covered.

In Chap. 3, various methods of reducing crosstalk in CNT interconnects are studied. First, the role of semiconducting CNTs in crosstalk reduction is discussed. The tentative flow for growing and integrating semiconducting CNTs is shown.

Then, the case of mixed CNT bundles surrounded by semiconducting CNTs is analysed. The idea of rearranging SWCNTs and MWCNTs in a mixed CNT bundle is considered. Two possible configurations are analysed. Their propagation delay and crosstalk-induced delay are found out and compared.

Next, geometry-based crosstalk reduction methods for CNT interconnects are discussed. Here triangular CNT bundles are proposed and modelled. Their performance in terms of coupling capacitance, propagation delay, crosstalk-induced delay and peak noise is presented.

Chapter 4 discusses the use of ultra-low-k dielectric materials for crosstalk reduction in CNT interconnects. As this is an emerging area, the dielectric materials that are recently synthesized such as boron nitride nanotubes, graphene oxide and its composites and metal organic frameworks are studied and their properties are enlisted. Next, small diameter CNTs are proposed as ultra-low-k dielectrics. Integration of these proposed ULK dielectric materials into CNT interconnects is a challenge. So, these challenges are outlined. The dielectric constants of small diameter CNTs for various CNT diameters are calculated from analytical expressions and are shown.

In Chap. 5, the possibility of using air as dielectric medium is discussed and CNT/air-gap interconnects are analysed. The coupling capacitance, RC delay and chip area of CNT/AG interconnects are calculated and compared with Cu/low-k interconnects.

In Chap. 6, the effects of temperature on the performance of CNT interconnects are analysed. The temperature coefficient of resistance (TCR) for SWCNTs is calculated, and the resulting R(T) values and R(T)C delay at different ambient temperatures are calculated. Next, CNTs are analysed as fractal media. The fractal dimension of a metallic CNT is modelled, and its values are calculated for different CNT diameters and lengths. Results show that the fractal dimension approaches to unity as the CNT length increases, indicating that heat flow will be continuous in long CNTs of length $L \geq 20Å$.

In Chap. 7, high-speed subthreshold CNT interconnects are modelled. CNTFET-based driver and load circuits are used to connect both ends of the CNT interconnect. Various configurations of inverter and transmission gate-based circuits are considered as drivers and loads. The channel length is optimized for maximum output subthreshold current. Next, various phonon scattering models are discussed that affect the resistance of the CNT interconnect. This includes acoustic phonon and optical phonon scattering, electron–phonon and phonon–phonon scatterings. Further, temperature effects on the scattering, and hence the mean-free path of electrons and phonons for subthreshold voltage-driven interconnects are discussed. Results show the temperature dependence on the RC delay analysis for nominal threshold voltage as well as subthreshold voltage-driven interconnects.

Vellore, India

P. Uma Sathyakam
Partha Sharathi Mallick

Contents

About the Authors

Dr. P. Uma Sathyakam is working as an Assistant Professor in the School of Electrical Engineering, Vellore Institute of Technology, Vellore, India since December 2012. He received his B.Sc. Electronics degree from the University of Kerala, Thiruvananthapuram in 2007, the M.Sc. degree in Applied Electronics from Bharathiar University, Coimbatore, in 2009, M.S. (by Research) in Nanoelectronics from VIT University in 2011 and the Ph.D. degree in Nanoelectronics from Vellore Institute of Technology in 2018. He was a Research Fellow in an MHRD sponsored project on 'online lab in microelectronics and VLSI' from July 2009 to July 2010 in VIT and was a Research Associate from August 2010 to June 2011 during his M.S. in VIT. His areas of interest are nanoscale interconnects, carbon nanotube electronics, graphene and novel electronic materials, and futuristic electron devices. He published 18 journal and conference papers of international repute and a book on 'VHDL Implementation of a 16-bit microprocessor'. He is a senior member of IEEE.

Dr. Partha Sharathi Mallick is a Professor and former Dean of the School of Electrical Engineering, Vellore Institute of Technology, Vellore, India. He is also the Director of Office of the Ranking and Accreditation, VIT. He was the Technical Head of IAAB Electronics a Swedish Industry in Bangladesh. He led various research teams and developed online laboratory in microelectronics, Monte Carlo simulator of compound semiconductors, nanostructured MIM capacitor, and low-cost electric fencers. He has authored more than 100 research papers in different journals and conferences of international repute. He is the past Chapter Chair and a present Chapter Adviser of the IEEE-EDS VIT of Region 10 Asia-Pacific. He was an enlisted technical innovator of India in 2007. He has published a book on MATLAB and Simulink and in IET, UK, and has published his book chapter on MIM Capacitor in June 2016. His areas of interest are finding new materials and technology for future nano-scale electronics.

Chapter 1
Introduction

This chapter gives an introduction to the VLSI interconnect technology. Emerging interconnect technologies like carbon nanotube and graphene nanoribbon interconnects are overviewed. Of the two, the analysis and design of carbon nanotube interconnects is taken up in this book. The current interconnect requirements for VLSI are outlined that is essential for making new interconnect technologies a reality. Also, this chapter gives a broad outlook on various topics that are discussed in this book.

1.1 Interconnect Technology for VLSI

Transistors and interconnects are part and parcel of an integrated circuit (IC). Transistors perform various functions that an IC is designed for, while interconnects connect the transistors to carry various lines such as power, ground, logic signals and clock pulses. Based on the length of interconnects, they can be classified as local, intermediate and global interconnects.

Local interconnects are used to connect between various transistors and logic blocks. Their lengths can range from a few tens of nanometres to 100 μm. Intermediate wires are used to connect between two complex logic blocks (CLBs) or two cores. They can have lengths up to 500 or even 900 μm. Global interconnects carry power, ground and clock signals to all the logic cores in an IC. They are very long with lengths from 1000 μm and higher.

The sizes of the transistors and interconnects are scaled down as the technology progresses. Therefore, traditionally used aluminium interconnects are replaced with copper interconnects as they have higher conductivity and are down scalable. Introduction of Cu interconnects made it easy for CMOS devices to be scaled down to a few microns and later to even nanometres. Also, Cu interconnects enabled the use of dual damascene technology in fabrication of CMOS devices which increased the density of transistors in ICs.

P. U. Sathyakam and P. S. Mallick, *Design and Crosstalk Analysis in Carbon Nanotube Interconnects*, https://doi.org/10.1007/978-981-15-8888-4_1

At the turn of the twenty-first century, new architectures came into existence such as 3D ICs, system on chip (SoC) and system on package (SoP). Consequently, both horizontal and vertical interconnects were used. Vertical interconnects or *vias* are used to connect between two levels or from one level to the substrate in 3D ICs. As the device and interconnect density is increasing further down to nanometre scale, Cu can no more be the choice as interconnects. In the presently used 14 nm technology node devices, Cu wires are already facing reliability problems, e.g. high wire resistivity, grain boundary scattering, electromigration and thermal issues. Additional barrier, cladding and etch stop layers for Cu lining take up considerable of chip area and are not desirable for highly scaled-down chips.

1.2 Emerging Interconnect Technologies

Alternate interconnect technologies are emerging that can replace Cu interconnects in future ICs. Additionally, these technologies are more attractive as they are more suitable for integration in 3D ICs, SoCs and SoPs. Two interconnect technologies that are emerging in the past decade are discussed below.

1.2.1 Carbon Nanotube Interconnects

Carbon nanotubes are the most extensively studied material for interconnect applications ever since its discovery by Iijima and Ichihashi (1993). This is due to their excellent intrinsic properties that make them one of the ideal candidates as nanoscale interconnects. CNTs are found to have extraordinary physical and chemical properties, e.g. high tensile strength of 100 GPa, a very high Young's modulus of around 1 TPa, a high percentage of elongation at break—16%, super-hydrophobic with a contact angle of 160°, high electron density of 4×10^9 A/cm^2 and an extremely high thermal conductivity of 3500 W/mK. As a matter of fact, CNTs are one-dimensional nanowires which exhibit quantum wire behaviour with ballistic electron transport. The sp^2 hybridized covalent C-C bonds make CNTs the strongest material available. The electrons in $2p_Z$ π orbitals contribute to the electron conduction in CNTs. Obviously, due to their high surface-to-volume ratio, the electrons contributing to the electrical conduction are all localized at the surface of the nanotubes. For interconnect applications, single-walled CNT (SWCNT), multi-walled CNTs (MWCNTs) and mixed CNT bundles are proposed. Due to the high resistivity of a SWCNT, which is around 6.45 kΩ given by h/4e^2, bundles or ropes of SWCNTs are proposed as interconnects. MWCNTs, on the other hand, have random chirality of their shells. Statistically one-third of the shells in an N shell MWCNT are metallic while two-third are semiconducting. This book particularly details the present state-of-the-art performance analysis of CNT interconnects for VLSI circuits. More details concerning various models of CNT interconnects and their crosstalk and thermal analysis will be provided in next sections.

1.2.2 Graphene Nanoribbon Interconnects

Graphene is a well-known material for its exceptional physical and electrical properties similar to CNTs. It is called as a zero bandgap semiconductor as its valence band maxima and conduction band minima meet exactly at the Fermi level. When graphene is cut into ribbon form, graphene nanoribbons are formed. Both single-layer and multi-layer GNRs (MLGNRs) are studied as interconnects, Bhattacharya et al. (2016). GNRs can have armchair or zigzag arrangements of carbon atoms which render them either metallic or semiconducting, respectively. Therefore, armchair GNRs are proposed as interconnects.

GNRs have same conducting properties similar to metallic CNTs and their current density is of the same order of magnitude as that of CNTs Naeemi and Meindl (2007a). Apart from these, the only advantage that GNRs have over CNTs is that they are easy to fabricate and integrate into ICs and as in the case of CNTs, chirality control is needed.

However, one of the main limitations of GNRs is the edge scattering of electrons. As GNRs have open edges on both sides, electrons scatter out at the edges and thus increase its resistance. Hydrogen linking of the open carbon bonds along the edges of GNRs is reported as discussed by Naeemi and Meindl (2007a). But, if the scatterings at the edges are specular, there will not be much difference in the GNRs conductance. Specular scatterings occur only when the roughness of the boundary is much smaller than the Fermi wavelength F = $2/k_F$. Typically, graphene has a wavelength of 1 nm, which means smooth edges are needed for specular scatterings to occur. The conductance of a GNR is found to be affected by its width (Naeemi and Meindl 2007a). Wide GNRs (W > 100 nm), both metallic and semiconducting, are having similar conductance. As the width reduces, its resistance increases. At around 8 nm width, metallic GNRs were found to outperform Cu wires in terms of conductance. Also, monolayer SWCNT interconnects were found to be a better choice than GNRs as they have larger resistance compared to the former case (Naeemi and Meindl 2007a). Nevertheless, many researchers worked on the modelling and performance analysis of GNR interconnects. Nasiri et al. (2010) showed that the stability of GNRs depends on its length and width.

GNRs have been fabricated for interconnect applications. Jiang et al. (2017) fabricated GNRs using chemical vapour deposition (CVD) and by intercalation doping by $FeCl_3$. Analysis of fabricated structures showed that due to the efficient diffusion of $FeCl_3$ into graphene, they can outperform Cu interconnects below 20 nm nodes. Further, the current density of such doped MLGNRs was found to be 2×10^8 A/cm^2 experimentally (Jiang et al. 2017), which is far greater than the current density in Cu interconnects of the same dimensions. The performance of MLGNR interconnects depends on the interlayer tunnelling resistance Rai et al. (2016). The tunnelling resistance depends on the length and the width of the GNRs apart from the number of layers in a MLGNR. Temperature-dependent performance analysis of GNR interconnects was carried out by Rai et al. (2017). Most of the studies on GNRs as interconnects show their performance benefits over Cu interconnects.

1.3 Current Interconnect Requirements

The major factors that influence the performance of present and future interconnects for VLSI circuits are (i) RC delay or propagation delay, (ii) crosstalk, (iii) noise, (iv) power dissipation and (v) thermal stability. The main reliability challenges of present interconnects are summarized in Table 1.1.

Firstly, the RC delay of the interconnect is dependent on its geometry and contacts with the gate metal. The overall resistance is dependent on the wire resistance and the contact resistance. To decrease the propagation delay of a wire, resistance must be kept to minimum. The resistance of a wire increases as its dimensions are reduced. Consequently, at highly scaled-down technologies, wire resistance causes a major problem. On the other hand, the wire capacitance reduces as the wire size decreases which is desirable. Thus, to keep the RC delay to a minimum, the wire dimensions must be chosen such that the resistance and the capacitance are less. The next big issue of scaled technologies is that the distance between adjacent wires can become very small such that crosstalk arises between them. In other words, the coupling capacitance between the adjacent wires increases as the separation decreases. This capacitance can delay the signal propagation in wires and can even lead to logical errors. This delay is called as crosstalk-induced delay. Further, this can lead to noise in signals that are propagating in wires due to crosstalk. So, in order to reduce crosstalk-induced delay and noise, ultra-low-k (ULK) dielectric materials must be used to shield the wires from crosstalk. Because of their lower dielectric constant, the coupling capacitance between the wires can also be reduced. Next, the power dissipated by the wires can be due to two reasons. One is due to the intrinsic resistance of the interconnects which arises as explained above. This power dissipation is static in nature and can be reduced by lowering the resistance of the wire. The second reason is due to the switching activity of the interconnect and also due to the supply voltage and can be expressed as

$$P = \sum AFj.Cj.V^2.f \tag{1.1}$$

Table 1.1 Reliability issues and future requirements of CNT interconnects

Reliability concerns	Future requirements
Thermal bottlenecks	New thermal interface materials (TIM) for heat dissipation
RC delay	Driver/load sizing; Aspect ratio (AR) optimization for reduced delay
Crosstalk	Alternate wire cross-sectional geometries; New low-k materials-effective k-value < 2
Low-k materials	Resistance to oxidation, good adhesion, high tensile strength, chemical stablility, thermal stability, no porosity, hydrophobic behaviour

where AF_j is the activity factor, C_j is the capacitance of the wire, V is the supply voltage and f is the frequency of operation. This power is called as the dynamic power and is the main contributor of power dissipated in a wire. This dissipated power increases linearly with the frequency of operation. However, the increased power dissipation can be countered by decreasing the power supply which is the current trend in VLSI.

Lastly, the thermal stability of the interconnects is one of the most important factors that influence interconnect performance and reliability. Due to high-frequency switching, both transistors and interconnects can get heated up and increase the ambient temperature inside ICs. The operating dynamics of ICs are affected at elevated temperatures. Mainly, the interconnects must cope up with this heat as their resistance increases. Secondly, the current conduction is now coupled with heat conduction where scattering mechanisms come into play. Therefore, the thermal conductivity of the interconnect is very important. If the phonon scattering decreases, thermal conductivity will increase and hence the signal transmission will improve. Unfortunately, Cu interconnects are facing problems in all these issues at highly scaled-down technologies. Fortunately, carbon nanotubes are the materials which can outperform Cu wires in all respects. Next section explains about CNTs as the ideal candidates for VLSI interconnects.

1.4 Motivation and Objectives of This Book

In view of the limitations of copper interconnects in integrated circuits, carbon nanotubes are proposed as the alternative interconnect material. Table 1.1 shows the summary of properties of CNTs that make them ideal candidates as future VLSI interconnects (Bellucci 2005; Chae and Kumar 2006; Meo and Rossi 2006; Sinnott and Andrews 2001; Demczyk et al. 2002).

It can be seen from Table 1.2 that CNTs are superior to copper in electrical and thermal conductivities, are more mechanically stable and can conduct current at frequencies beyond >2 GHz. So, the main motivation of this book is to explore the use of CNTs as next-generation interconnects for nanoelectronic ICs. CNTs have already been studied as interconnects in the past decade.

Banerjee and Srivastava (2006), and Srivastava and Banerjee (2005) showed that due to the smaller resistivity of bundles of SWCNTs and MWCNTs, they can replace Cu wires in future ICs. Other research teams including Naeemi and Meindl (2005b), Naeemi et al. (2005), Naeemi and Meindl (2007b), Srivastava et al. (2009), Ceyhan and Naeemi (2013), Raychowdhury and Roy (2006), Nieuwoudt and Massoud (2006) worked on the modelling and simulation of CNT interconnects. From the analysis of CNTs as interconnects through modelling and simulations, it was found that they suffer from crosstalk due to high coupling capacitance. Crosstalk modelling was carried out by Rossi et al. (2007), and it was found that by reducing the coupling capacitance between adjacent CNT interconnects, the resulting crosstalk can also be reduced. There are many methods to reduce the crosstalk like

Table 1.2 Summary of properties of carbon nanotubes

Property	Value	Reference
Density	1.33 gm/cm^3	Chae and Kumar (2006)
Bandgap	0 eV for Conducting, 0.5 eV for Semiconducting	Bellucci (2005)
Current density	10^{13} A/cm^3	Bellucci (2005)
Thermal conductivity	2000 W/m K	Bellucci (2005)
Phonon mean free path	1000 nm	Bellucci (2005)
Relaxation time	10^{-11} s	Sinnott and Andrews (2001)
Young's modulus	1 TPa	Meo and Rossi (2006), Demczyk et al. (2002)
Tensile strength	30 GPa	Meo and Rossi (2006), Demczyk et al. (2002)

- using dielectric materials with smaller dielectric constants,
- increasing the spacing between the interconnects and
- reducing the height of the interconnects.

Using ultra-low-k (ULK) materials is a good idea. There are emerging ultra-low-k materials based on polymers that can be exploited for use in CNT interconnects. However, the next two conventional methods are not suitable for futuristic scaled-down technologies. So, alternate means for reducing crosstalk must be found out. Further, the thermal stability of CNT interconnects needs to be analysed in depth and their electrical conductivity mechanisms at high temperatures must be understood. Scattering of electrons and phonons occurs at high temperatures. Even though CNTs are good thermal conductors, the carriers (electrons and phonons) are confined to one dimension only and the mean free path of the carriers decreases as the ambient temperature increases. The complex interplay between the ambient temperature and the change in resistance of CNTs is not understood well. Moreover, CNTs are discrete structures. Heat flow through CNTs was modelled earlier using the heat diffusion equation and cannot predict the actual heat transfer phenomena in CNTs and bundles of CNTs. New heat conduction models that consider CNTs as discrete media must be developed to capture the heat flow in such media. The main objectives of this book are

- to identify new methods and materials for reducing crosstalk in CNT interconnects,
- to study and analyse temperature-dependent performance of CNT interconnects and
- to study and analyse new models of heat flow in CNTs where CNTs are treated as discrete media.

CNT-based interconnects can become the ultimate choice for next-generation IC technologies if the above points are addressed. This book takes a step further by doing in-depth analysis by proposing alternate methods for reducing crosstalk-induced delay and noise in CNT interconnects. Further, new thermal models are proposed here that can solve the complex heat flow mechanism in CNT interconnects. The next section describes the main contributions of this book that can be used to solve the problems of CNT interconnects.

1.5 Main Contributions

Small diameter semiconducting CNTs are proposed as low-k dielectric materials for the first time. Based on their radial polarizability calculations, their dielectric constant is found out. There is a direct correlation between the radius of the CNTs and the dielectric polarizability. As radius reduces, polarizability also reduces due to decreased dipole moment. Hence, the dielectric constant also decreases. In order to incorporate semiconducting CNTs in CNT interconnects, a novel CNT–metal contact geometry is proposed. This contact geometry helps in growing semiconducting small diameter CNTs at the periphery of the contact and metallic CNTs at the centre. Analysis of these bundles and similar types of bundles, where both SWCNTs and DWCNTs at the centre and semiconducting CNTs at periphery of the bundle, shows that the coupling capacitance can be reduced drastically and hence the crosstalk and induced delay and peak noise can also be reduced substantially. This benefit of reduction in crosstalk was studied for all possible switching scenarios of two coupled CNT bundle interconnects.

The next contribution of this book is to use a new type of CNT bundle cross-sectional geometry to minimize crosstalk between adjacent CNT interconnects. Among many geometries like square, rectangular, hexagonal, triangular and so on, the most optimal geometry that can reduce the coupling capacitance between adjacent CNT bundle interconnects is the triangular one. Transmission line modelling of triangular CNT bundles is carried out and the corresponding equivalent single conductor (ESC) parameters are calculated. Then, transient SPICE analysis using Silvaco SmartSPICE is carried out to verify and compare the propagation delay, power dissipated and crosstalk-induced delay with traditionally used square CNT bundle interconnects.

A comprehensive review of the emerging ultra-low-k dielectric materials that are possible candidates as shielding material for CNT interconnects is carried out. The materials studied include boron nitride nanotubes, graphene oxide/polyimides and metal organic frameworks apart from small diameter CNTs. Furthermore, various possibilities of integration of these materials into CNT interconnects and challenges that can arise during integration process have been identified. This enables researchers and engineers to make judgements based on the pros and cons of these materials to integrate them in future IC technologies.

The design and analysis of CNT interconnects with air gaps is done next in this book. As air has a relative dielectric constant of unity, they can be the most ideal low-k dielectric medium between CNT interconnects. This can benefit in reducing the coupling capacitance and hence crosstalk-induced delay and power dissipated by the interconnects. Further, the chip area consumed by the CNT/air-gap interconnects can also be reduced compared to copper/low-k interconnects which is advantageous for scaled-down ICs.

Next, the thermal analysis of CNT interconnects is done. First, the thermal stability of CNT/air-gap interconnects is analysed. The temperature coefficient of resistance is used to model the thermal behaviour of CNTs. The ambient temperature plays a major role in the conductivity of CNTs. The resistance of a CNT and its mean free path was found to be affected by the ambient temperature.

The thermal models that are used to describe heat flow in CNTs are based on continuum approach. In order to capture the actual heat conduction mechanisms of CNTs, they need to be treated as discrete media. This kind of discrete media are called as Fractals. So, CNTs are analysed as fractals and their fractal dimensions are modelled and calculated. Fractal dimensions of armchair SWCNTs are modelled and calculated. Knowing the fractal dimension enables one to solve the fractional differential equations that are applied to analyse the heat flow in CNTs. The values of fractal dimension for different lengths and diameters suggests that CNTs are good thermal conductors laterally but are bad conductors radially.

Further, temperature-dependent analysis of CNT interconnects when subthreshold current flows through them is studied. The mean free path of CNTs was profoundly dependent on temperature. Various scattering mechanisms dominate the current flow at elevated temperatures. Principal scattering mechanisms are due to acoustic phonons and optical phonons (Rai et al. 2017). The resistance of a CNT, which is dependent on the mean free path, is governed by these scattering mechanisms. Hence, the acoustic phonon and optical phonon mean free paths are modelled and calculated which shows the dominance of optical phonon mean free path of 1.6 μm compared to the smaller mean free path of around 10 nm for acoustic phonons. As temperature increases, the resistance and quantum capacitance of CNT interconnects increase and the kinetic inductance decreases.

Chapter 2
Background

This chapter gives an introduction of carbon nanotube (CNT) interconnect technology in detail. First, recent developments since the past decade are discussed elaborately. CNT interconnect models like Luttinger liquid theory and transmission line models are outlined. Then, various aspects of crosstalk analysis of CNT interconnects are discussed. Factors that affect crosstalk and the ways to reduce crosstalk are also outlined.

2.1 Introduction

The use of integrated circuits (ICs) in the electronics industry has propelled its growth in the past four decades. Today, majority of the electronics applications and gadgets utilize integrated circuits. Over the years, researchers have recognized that miniaturization of the components in an IC, i.e. transistors and interconnects, can increase its speed and reduce the power consumption.

Extensive research was started by Intel Corporation led by scientist Gordon Moore in 1965, who predicted that the number of transistors in an IC will double in every 18 to 24 months. In order to meet this target, researchers started inventing new transistors made of novel materials. From the traditional silicon MOSFET, new materials such as GaAs and other compound semiconductor-based high electron mobility transistors (HEMTs), FinFETs, Si nanowire FETs, CNTFETs, GNRFETs and now organic FETs (OFETs) are invented. These novel devices are ultra-small with dimensions ranging from a few nanometres to a few tens of nanometres. These devices reduce the power consumption and increase the speed of operation in addition to size reduction.

Simultaneously, interconnects connecting the transistors are studied. Aluminium interconnects were used to drive the ICs initially. After the introduction of complementary metal oxide semiconductor (CMOS) technology, Al was replaced with

P. U. Sathyakam and P. S. Mallick, *Design and Crosstalk Analysis in Carbon Nanotube Interconnects*, https://doi.org/10.1007/978-981-15-8888-4_2

copper due to its comparatively smaller resistivity when it is scaled down. Introduction of copper interconnects was followed by the dual damascene technology where copper is melted and poured into Si trenches. To protect Cu from diffusion into Si, a cladding and an etch stop layer is coated upon Cu. This makes Cu wires more mechanically stable. Initially, the performance of the IC mainly depended on transistors. However, starting the microelectronics era and using scaled transistors, the performance of the interconnects became the focused area of research. The International Technology Roadmap for Semiconductors (ITRS) was constituted in 2001. Prior to the then NTRS, National Technology Roadmap for Semiconductors in 1994 projected the future requirements of new materials, e.g. Al and Cu as interconnects and the need for new dielectric materials. The NTRS is reconstituted as ITRS in 2001. The ITRS committee convened binally and charted out the new requirements and challenges for the IC industry (http://www.itrs2.net/itrs-reports.html). In its final report on interconnects in 2015, it described the interconnects at the forefront of challenges in microelectronics industry. The five most difficult challenges that the industry face for dimensions less than 16 nm are materials, metrology, process, complexity of integration and practical approaches for 3D.

Figure 2.1 shows the cross-section schematic representation of a multi-layer IC with Cu interconnects. The pattern of Cu wires is carried out by dual damascene

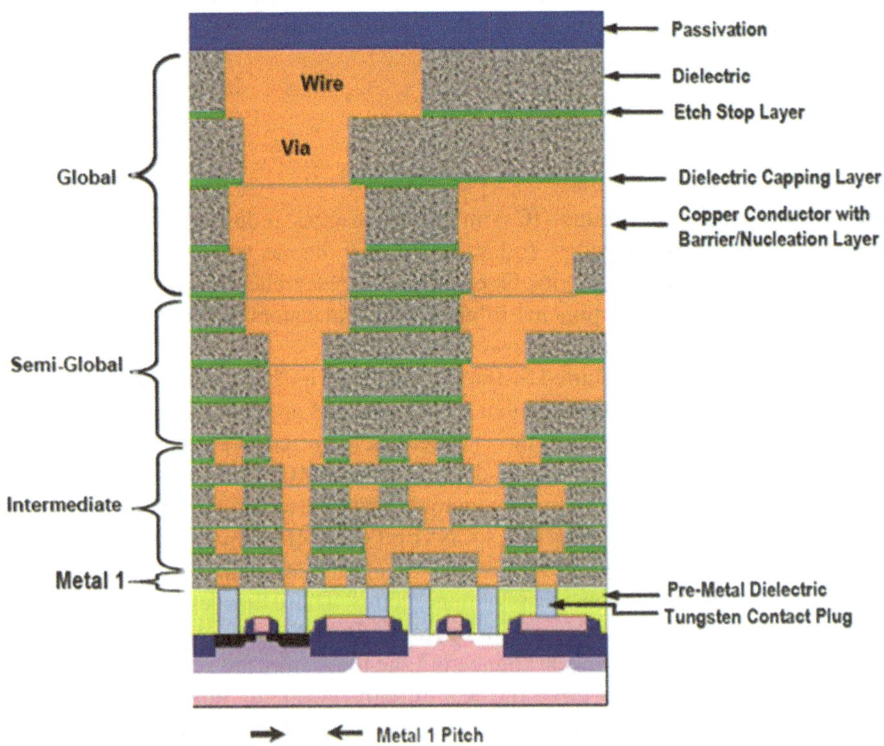

Fig. 2.1 Cross section of a dual damascene processed multi-layer IC (ITRS-2015)

process. This technique was adopted when the interconnect material was switched to Cu from Al as plasma etching cannot be performed for Cu interconnects. A detailed IC process flow is described in the ITRS (2015) report (http://www.itrs2.net/itrs-reports.html). The technology node at which devices are manufactured is an important parameter. From 0.1 μm technology node devices that used Al interconnects, the current node size stands at 12 nm where FinFETs and Cu interconnects are used. Beyond this node size, Cu cannot be used due to reliability problems. The Cu resistivity increases due to its extremely reduced size, crosstalk and power dissipation increase, electromigration can break the wires and thermal integrity is a question. Therefore, alternative materials such as CNTs and GNRs are proposed for the realization of the next-generation interconnects. Table 2.1 shows the list of contributions to the literature on the fundamental modelling of carbon nanotubes as interconnects.

2.2 Carbon Nanotube Interconnect Models

2.2.1 Luttinger Liquid Theory

Bockrath et al. (1999) had first proposed the Luttinger liquid theory for electron transport in carbon nanotubes. In one-dimensional system like CNTs, even small Coulomb interactions can cause strong perturbations. Consequently, to understand the conducting properties of CNTs and the resulting electron transport, experiments were conducted by growing CNTs and connecting the ends to contact metals (Bockrath et al. 1999). The current flow through the CNT ropes was studied with respect to change in the gate voltage at various temperatures as shown in Fig. 2.2. In the top contact method where chromium-gold contacts are deposited on the nanotube rope ends, the electron transport is mainly due to tunnelling of electrons through the ends. But, in the second method, where the leads are deposited on the substrate first and CNT ropes are deposited on top of the leads, electrons travel by tunnelling in to the middle of the nanotube. Samples in the second case showed Coulomb blockade behaviour at low temperatures which is analogous to a single quantum dot at the ends. Therefore, the CNTs are referred to as the bulk contacted CNTs.

From Fig. 2.2, the Coulomb oscillations are clearly visible when electrons are added to the nanotube at low temperatures. However, as the charging energy U, which is equal to 1.9 meV for CNT ropes used here, exceeds the thermal energy (K_{BT}) above 20 K, the conductance becomes independent of the gate voltage. An in-depth analysis of the two cases of contact geometry is carried out by Bockrath et al. (1999). This model is further extended by Burke (2002). He first proposed the models for interacting electrons and spin that can help in formulating the transmission line RLC parameters as shown in Fig. 2.3a and b. Experimental measurements of 2D plasmons were performed by Burke (2002). It was suggested that the same mechanism can be used to observe 1D plasmons in SWCNTs. Frequency-domain measurements suggest that SWCNTs can operate at THz frequencies as well.

Table 2.1 List of publications on modelling of CNT interconnects

Reference	Work
Sanvito et al. (2000)	Fractional quantum conductance
Naeemi and Meindl (2005b)	Monolayer CNT interconnects
Srivastava and Banerjee (2005)	Circuit models and performance analysis of SWCNT interconnects
Naeemi et al. (2005)	Comparison of CNT and Cu interconnects
Nihei et al. (2005)	Parallel channel conduction in MWCNTs
Li et al. (2005)	Multichannel ballistic transport in MWCNTs
Banerjee and Srivastava (2006)	Circuit model of SWCNT bundles
Raychowdhury and Roy (2006)	Circuit model of SWCNT bundles and comparison with Cu
Nieuwoudt and Massoud (2006)	Diameter dependent resistance models
Naeemi and Meindl (2006)	Physical models of MWCNTs
Naeemi and Meindl (2007b)	CNTs as Local, Semi-global, and Global interconnects
Li et al. (2008)	Circuit modelling and performance analysis of MWCNT interconnects
Li and Banerjee (2009)	High frequency analysis of CNT interconnects
Srivastava et al. (2009)	Circuit models and temperature dependent performance of SWCNT interconnects
Li et al. (2009)	CNT interconnects review
Fathi et al. (2009)	Transmission line models of CNT wires
Fathi and Forouzandeh (2009a)	Stability analysis of CNT interconnects
Fathi and Forouzandeh (2009b)	Time domain analysis of distribute RLC models of CNT wires
Xu et al. (2010)	Waveguide like behaviour in MWCNTs
Sarto and Tamburrano (2010)	Frequency dependent TL model of MWCNTs
D'amore et al. (2010)	ESC and MTL models of SW/MW CNT interconnects
Li et al. (2011)	CNT vias-performance analysis
Liang, Wang and Ding (2011)	Repeater insertion in MWCNT interconnects
Moghaddam et al. (2013)	SWCNT inductance behaviour
Ceyhan and Naeemi (2013)	Single and few SWCNT interconnects
Ferranti et al. (2014)	Parametric macromodels for MWCNT interconnects
Kumar et al. (2015)	Stability and delay analysis of MWCNT interconnects
Tang and Mao (2015)	Fast simulations of ESC and MCC models of MWCNT interconnects

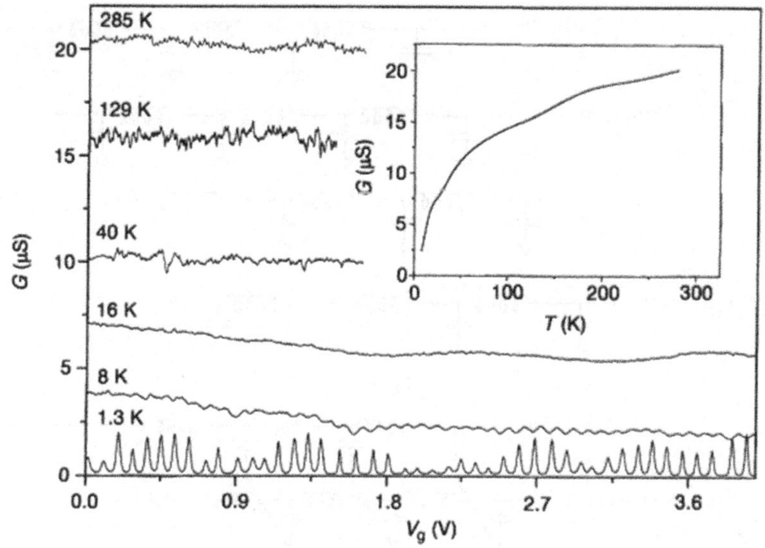

Fig. 2.2 Conductance versus gate voltage of CNT ropes at various temperatures. (Bockrath et al. 1999)

2.2.2 *Electron Transport Model*

In order to understand the electrical conduction behaviour of CNTs, they are modelled as transmission lines (TL) which takes into consideration the electron transport in CNTs. Therefore, metallic CNTs are considered as conductors that are placed on a ground plane or substrate as shown in Fig. 2.4. Here, d is the diameter of the CNT and h is the height at which the CNT is placed from the ground plane. The conductance of a CNT can be simply given as e^2/h, where h is Planck's constant and e is the charge of electrons. Thus, the resistivity per one micrometre length of CNT is h/e^2. Considering the spin of electrons, i.e. spin up and spin down, and the number of conducting channels, which are two for a 1 nm diameter CNT, the total resistance is $h/4e^2$ which equals to 6.45 kΩ/μm. This resistance is called as the quantum resistance (R_q) which is the fundamental and intrinsic resistance of a CNT. The total amount of energy stored in a CNT can be given by the magnetic and kinetic inductances. The magnetic inductance of a CNT is calculated by equating the inductive energy to the stored magnetic energy

$$\frac{1}{2}LI^2 = \frac{1}{2\mu}\int B(x)^2 dx \tag{2.1}$$

which is the equation for CNT placed above a ground plane. Thus, the magnetic inductance is

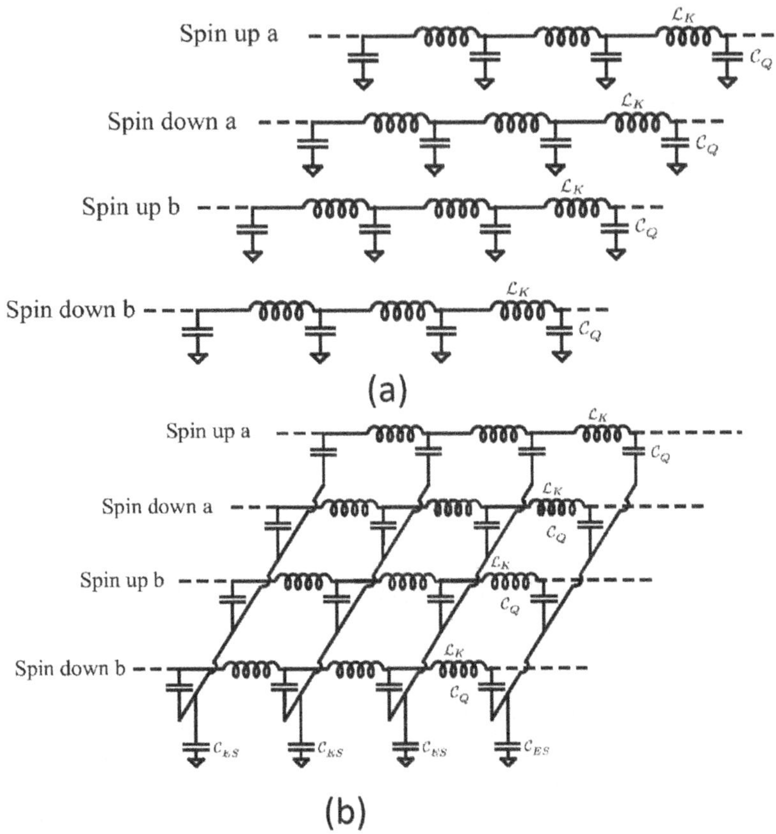

Fig. 2.3 **a** Non-interacting and **b** interacting electron model of a single-walled carbon nanotube. (Burke 2002)

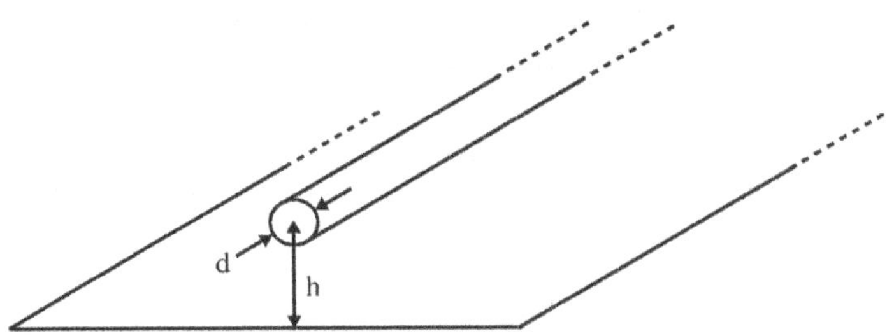

Fig. 2.4 A simplified model for the carbon nanotube placed above a ground plane

$$L_M = \frac{\mu}{2\pi} cosh^{-1}\left(\frac{2h}{d}\right) = \frac{\mu}{2\pi} \ln\left(\frac{h}{d}\right) \tag{2.2}$$

If d is taken as 1 nm and h as 1 m, the value of L_M can become 1 pH/μm. Next, the kinetic inductance arises due to the kinetic energy of electrons in the CNT. The derivation of this inductance can be found in Burke (2002). It can be given by

$$L_K = \frac{h}{2e^2 v_f} \tag{2.3}$$

By substituting for h, e and the Fermi velocity v_f ($v_f = 8 \times 10^8$ m/s for graphene and CNTs), $L_K = 16$ nH/μm. Consequently, compared to L_M, L_K is greater by four orders of magnitude. Hence, in the following chapters, CNT analysis is carried out by considering L_K only. The total inductance is

$$L_{total} = L_K + L_M \tag{2.4}$$

Electrostatic capacitance arises between the CNT and the ground as CNT is carrying current. The capacitance is

$$C_E = \frac{2\pi\epsilon}{\ln(\frac{h}{d})} \tag{2.5}$$

and is 50 aF/μm considering the standard values of d = 1 nm and h = 1 μm and $\epsilon = 3.9$. Energy is required to add new electrons to the available states at the Fermi level (E_F). Due to the density of states available at E_F for a 1D quantum wire, which is expressed as the quantum capacitance. The capacitance is

$$C_Q = \frac{2e^2}{h v_f} \tag{2.6}$$

which can be calculated as 100 aF/m. The magnitudes of both electrostatic and quantum capacitances are comparable, and hence both must be considered in the circuit analysis of CNTs. So the total capacitance is

$$\frac{1}{C_{total}} = \frac{1}{C_E} + \frac{1}{C_Q} \tag{2.7}$$

The circuit model that incorporates the above elements can be drawn as shown in Fig. 2.5. This circuit is used to model CNT interconnects made of SWCNTs and its bundles, MWCNTs and mixed CNT bundles. A more detailed analysis and review of electron transport in SWCNTs is given in the book chapter by Biercuk et al. (2007).

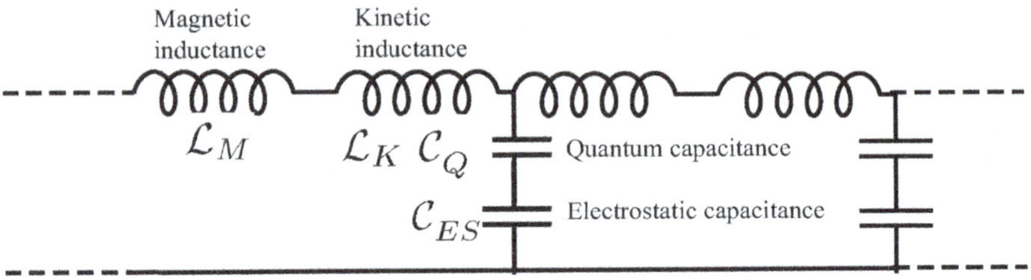

Fig. 2.5 Circuit model of a 1D CNT quantum wire

2.3 Review of Carbon Nanotube Interconnects

2.3.1 SW, MW and Mixed CNT Bundle Interconnects

Single-walled carbon nanotubes are the first type of CNTs that are studied as VLSI interconnects. Basically, there are two types of SWCNTs that are used for nanoelectronic applications (see Fig. 2.6). The armchair-type CNTs are metallic in nature and are used as interconnects. The zigzag CNTs are semiconducting in nature, and hence are used in CNTFETs. SWCNTs have very high quantum resistance of $6.45\,k\Omega$. Thus, they cannot be used directly as interconnects. Instead, bundles of CNTs are proposed as interconnects to replace copper as the equivalent resistance of CNT bundles is reduced and less than copper wires. Many researchers studied the behaviour of SWCNT bundle interconnects. Raychowdhury and Roy (2006) studied the performance of single SWCNTs against Cu interconnects. They observed that as single SWCNTs have high quantum capacitance, bundles of SWCNTs are needed to outperform the resistivity effects of Cu interconnects. Banerjee and Srivastava (2006) first proposed the equivalent circuit models for SWCNT bundle interconnects. Figure 2.7 depicts the schematic of a SWCNT bundle placed on a grounded substrate. Square-type CNT bundles (height/width aspect ratio (AR) = 1) were found to be better structures than rectangular ones (AR = 1.5, 2) as the former has lesser number of CNTs in adjacent bundles coupled to each other. The bundle is placed at a distance of

Fig. 2.6 Two types of SWCNTs

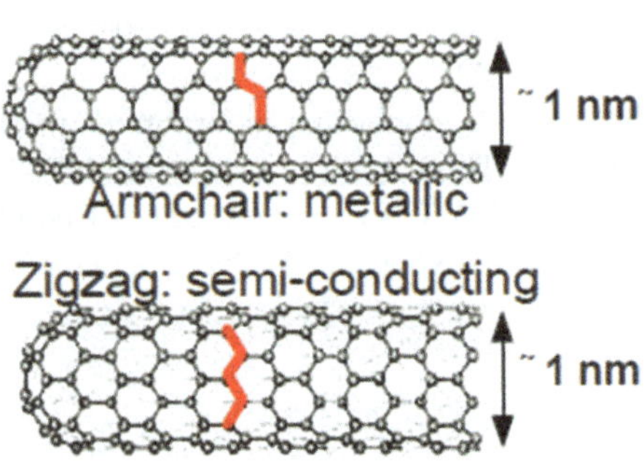

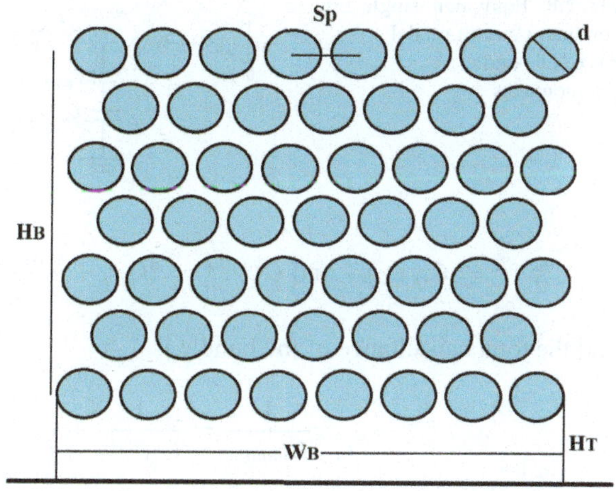

Fig. 2.7 Schematic of a SWCNT bundle placed on a grounded substrate

h_t from the ground plane. The number of CNTs along the width (W_B) of the bundle is n_W and the number of CNTs along the height (H_B) of the bundle is n_H. Each CNT has a diameter d of 1 nm. Then the total number of CNTs can be found as

$$n_{CNT} = \begin{cases} n_W n_H - (n_H/2), & \text{if } n_H \text{ is even} \\ n_W n_H - [(n_H - 1)]/2, & \text{if } n_H \text{ is odd} \end{cases} \tag{2.8}$$

where $n_C NT$ is the total number of CNTs in the bundle and

$$n_W = \left\lfloor \frac{W_B - d}{S_p} \right\rfloor \text{ and } n_H = \left\lfloor \frac{H_B - d}{S_p} \right\rfloor \tag{2.9}$$

where S_p is the centre-to-centre distance of adjacent tubes and is 1.34 nm in this case, considering the van der Waals gap δ of 0.34 nm. The total resistance of the SWCNT bundle is given as

$$R_{bundle} = \frac{R_{CNT}}{n_{CNT}} \tag{2.10}$$

where $R_{CNT} = R_Q \frac{l}{l_{mfp}}$, l is the length of the CNT and l_{mfp} is the mean free path of a CNT (=1 μm). The inductance of a CNT bundle can be given as

$$L_{bundle} = \frac{L_{CNT}}{n_{CNT}} \tag{2.11}$$

where L_{CNT} is the sum of the kinetic and magnetic inductances as given in Eq. 2.4. Also, the total quantum and electrostatic capacitances of a bundle are given as

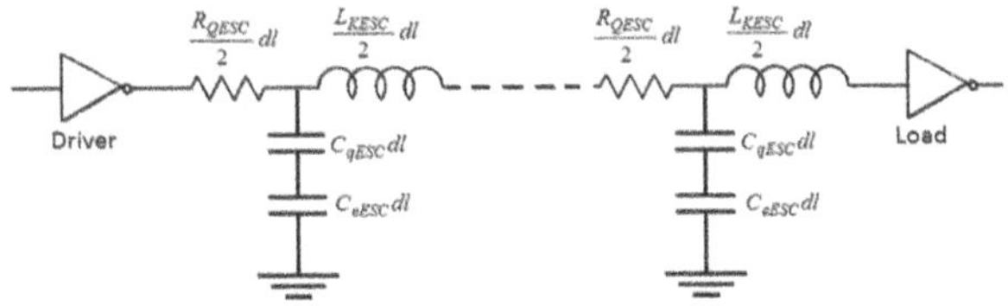

Fig. 2.8 Equivalent single conductor circuit model SWCNT bundle interconnects

$$C_Q^{bundle} = C_Q.n_{CNT} \text{ and } C_E^{bundle} = 2C_{En} + \frac{n_W - 2}{2}C_{Ef} + \frac{3(n_H - 2)}{5}C_{En} \quad (2.12)$$

and the total capacitance of the bundle is

$$\frac{1}{C_{bundle}} = \frac{1}{C_Q^{bundle}} + \frac{1}{C_E^{bundle}} \quad (2.13)$$

These equations are collectively called the equivalent single conductor (ESC) model of CNT bundle interconnects. The ESC model is used to analyse the performance of CNT interconnects with respect to Cu interconnects. The resulting equivalent single conductor circuit diagram is given in Fig. 2.8. The interconnect has lumped circuit elements as well as distributed circuit elements. The distributed R and L elements are divided between the two ends of the wire and hence are divided by 2. The contact resistance is the lumped element. The interconnect is driven by an inverter and is terminated by another inverter. A more detailed analysis of various inductive behaviours of SWCNTs is done by Moghaddam et al. (2013). The two main points raised by Moghaddam et al. (2013) is that, firstly, the magnetic inductances of bundled CNTs and Cu wires are same. Secondly, the inductive behaviour of CNT interconnects at current and future operating frequencies vanishes, and hence need not be considered. In the earlier years, CNT interconnects were modelled by considering fixed diameter of CNTs say 1 nm. However, Nieuwoudt and Massoud (2006) proposed diameter-dependent modelling of SWCNTs. This paved the way for the better understanding of the SWCNT interconnect performance. As can be seen from Fig. 2.9, the resistance of the interconnect and the mean free path (MFP) are dependent on the CNT diameter.

The diameter dependence of the resistivity of a CNT bundle shows that, for dense and sparse bundles, the resistivity increases with diameter and is still lesser than 22-nm-wide Cu interconnects as shown in Fig. 2.10. Also, the ratio of diameter dependent to fixed resistance increases as the diameter increases. This is reflected in the difference in the contact resistances for both the cases as shown in Figs. 2.11 and 2.12. Performance analysis of SWCNT bundle interconnects against copper interconnects was done by Naeemi and Meindl (2007b), Li et al. (2008) and Li et al. (2009). In all the studies, the performance factors like propagation delay, power dissipation and the power-delay product or energy-delay product are investigated and compared against each other. It was found that, by adjusting the CNT bundle design parameters, they can be designed to outperform Cu interconnects at 22 nm

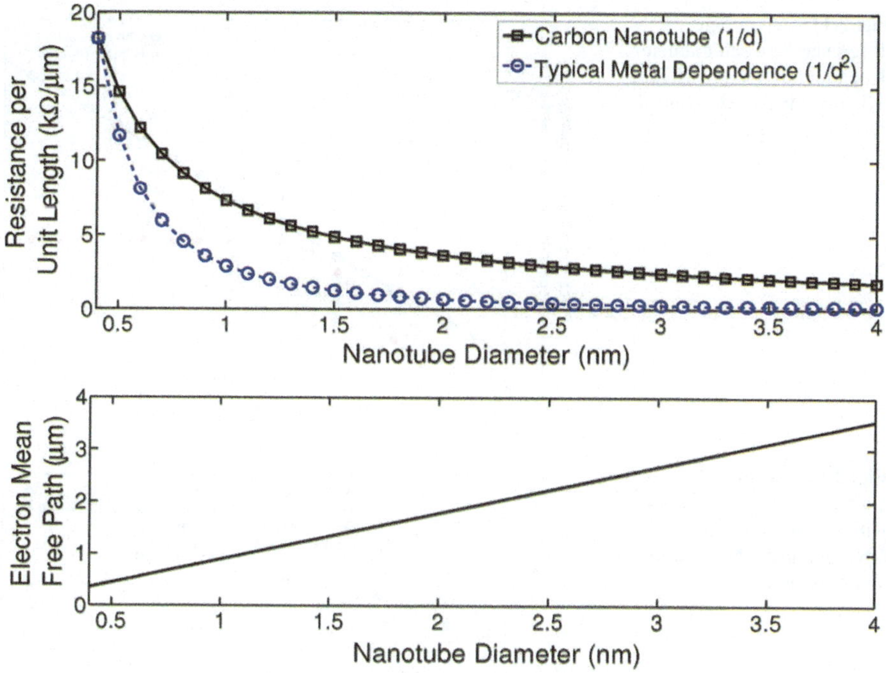

Fig. 2.9 Resistance and MFP variations with respect to CNT diameter (Nieuwoudt and Massoud 2006)

Fig. 2.10 Change of resistivity with respect to CNT diameter (Nieuwoudt and Massoud 2006)

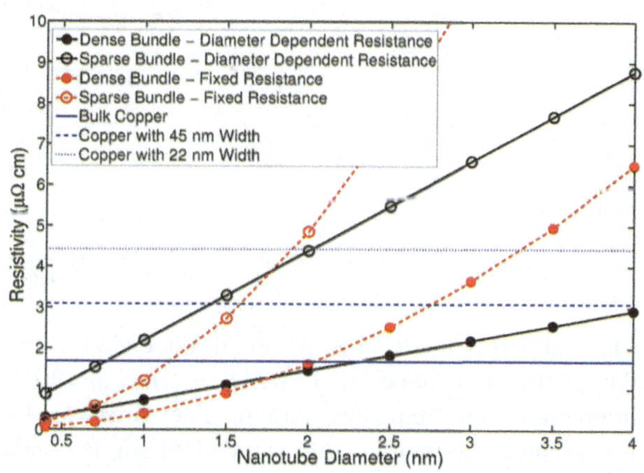

and below 22 nm technology nodes. A detailed encounter of CNT bundle design parameters is discussed in Sect. 2.4. Frequency-dependent modelling of SWCNTs was carried out by Li and Banerjee (2009). In this work, it is pointed out that due to significant skin effects and proximity effects, inductance must be considered in high-frequency analysis. The length-dependent impedance of a SWCNT interconnect can

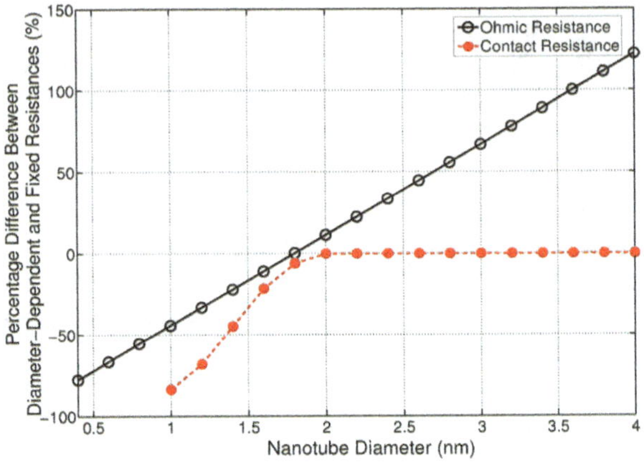

Fig. 2.11 Percentage difference between diameter dependent to fixed resistance (Nieuwoudt and Massoud 2006)

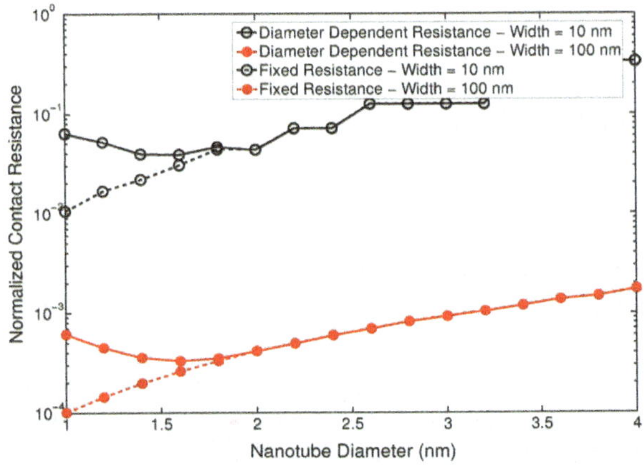

Fig. 2.12 Normalized contact resistance variations with respect to change in CNT diameter (Nieuwoudt and Massoud 2006)

be given as

$$Z_{pul} = \frac{h}{2e^2} \cdot \frac{1}{\lambda} + j\omega \frac{h}{4e^2 v_f} \tag{2.14}$$

The real part of the RHS is the length-dependent resistance and the imaginary part is the kinetic inductance L_K. Later D'amore et al. (2010) had studied SWCNT bundle interconnects in frequency domain. They proposed two models for SWCNT bundle interconnects. First one is the equivalent single conductor (ESC) already discussed above. This method considers the SWCNT bundle as a lumped single conductor by taking the series and parallel combinations of R, L and C as applicable. The second method is the multiconductor transmission line model (MTL) where each CNT in the bundle is treated individually and its R, L and C are calculated individually by formulating N×N matrices with N being the number of CNTs in the bundle. This approach can be considered when each CNT is excited by independent sources. In the case of SWCNT bundle interconnects, due to imperfect contacts, it was thought ideal

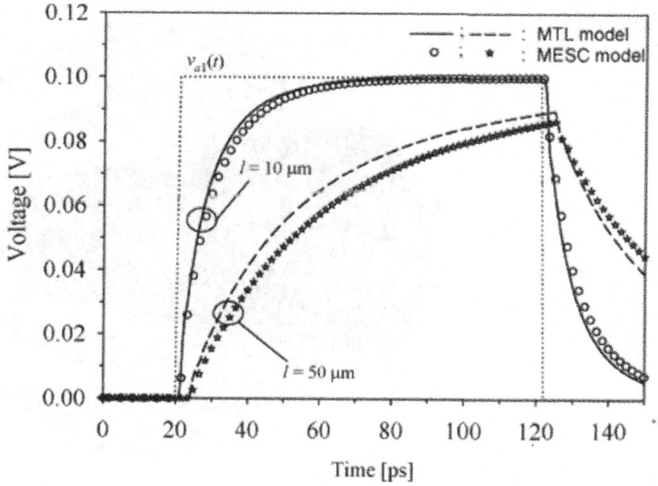

Fig. 2.13 Comparison of transient outputs of a SWCNT bundle interconnect at 14 nm technology node (D'amore et al. 2010)

to consider MTL model where each CNT is carrying a different voltage than the other in the bundle. The comparison of the accuracy of both the models shows that ESC model is more suitable for computation of the delay in CNT interconnects. Figure 2.13 shows the transient waveforms of both MESC and MTL models of SWCNT bundle interconnects at 14 nm technology node, for comparison. Recently, Ceyhan and Naeemi (2013) proposed using single or few CNTs as VLSI interconnects. They used ITRS predicted values of interlayer relative dielectric constant of 1.8 and showed that CNTs can replace Cu as local interconnects at 11 nm technology node and below. This is because the resistivity of local Cu wires is comparable to the intrinsic delay of MOSFET devices at 1 nm node size. Thus, leveraging on the advances of horizontal CNT wire fabrication technologies, the resistivity of short single or few CNTs will be less than Cu wires of the same size. The delay and power dissipated by such wires showed encouraging results when single or few SWCNTs were used as interconnects.

Most of the analyses of SWCNT bundle interconnects carried out till 2010 considered that statistically only one-third of the CNTs in the bundle are metallic while the rest two-third CNTs are semiconducting. However, due to matured fabrication techniques adopted to grow SWCNTs later works considered 80% and subsequently 100% metallic CNTs as done by Alizadeh and Sarvari (2016). A detailed analysis of crosstalk effects in SWCNT bundle interconnects is given in the next chapter.

MWCNT interconnects are more complex structures than SWCNT bundle interconnects. Nevertheless, MWCNT interconnects are given equal importance as they can perform better than Cu wires and even better than SWCNT bundle interconnects in some cases. Basically, MWCNTs are concentric cylinders of CNTs as depicted in Fig. 2.14. The number of concentric CNTs in an MWCNT can be from 2 to 10 CNTs. Their diameters can range from a few nanometres to a few tens of nanometres.

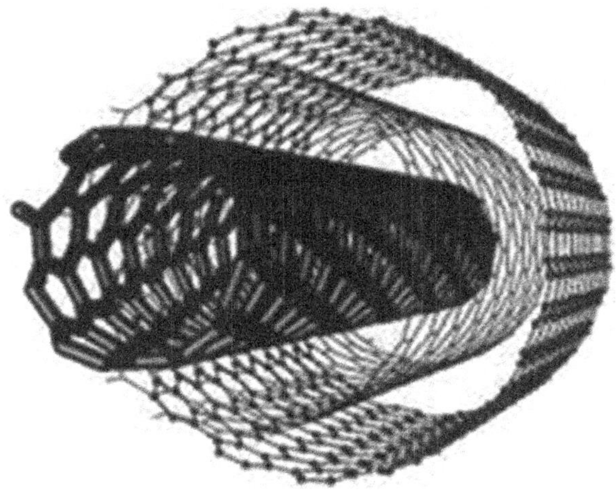

Fig. 2.14 Schematic of an MWCNT

MWCNT interconnects gained attention in 2005 when Li et al. (2005) showed the ballistic transport properties of MWCNTs. They experimentally proved that MWCNTs have very large current-carrying capacities up to 7.24 mA. The conductance and the resistance are measured as 460 to 490 S and 34.4 Ω, respectively. The main reason for the high conduction in MWCNTs is due to multichannel quasi-ballistic conducting behaviour of MWCNTs with large diameter. The metal contacts that were formed with the ends of MWCNTs are Ohmic in nature. Nihei et al. (2005) showed that vertically grown MWCNTs on Co catalyst and Ti contacts have a resistance of 0.7 KΩ. One of the main reasons of early fabrication and analysis of MWCNTs is that they are easy to grow than metallic SWCNTs. Modelling was done by Naeemi and Meindl (2006) by calculating the number of conducting channels per shell

$$N_{chan/shell} = \sum_{subbands} \frac{1}{exp(|E_v|K_B T) + 1} \tag{2.15}$$

This can be rewritten using the diameter (D) dependence of the number of conducting channels as

$$N_{chan/shell}(D) = \begin{cases} aD + b, & \text{if } D > 3\,\text{nm} \\ \frac{2}{3}, & \text{if } D < 6\,\text{nm} \end{cases} \tag{2.16}$$

where a is 0.0612 nm^{-1} and b is 0.425. Even though MWCNTs have semiconducting shells in it, electrons flowing through them will be thermally excited at higher temperatures of up to 90 °C. Hence, all the shells in an MWCNT, both metallic and semiconducting, contribute to the conductivity of the CNT. This is called multiple shell conduction in MWCNTs. Li et al. (2008) modelled MWCNTs as transmission lines as shown in Fig. 2.15. The number of shells in an MWCNT can be found as

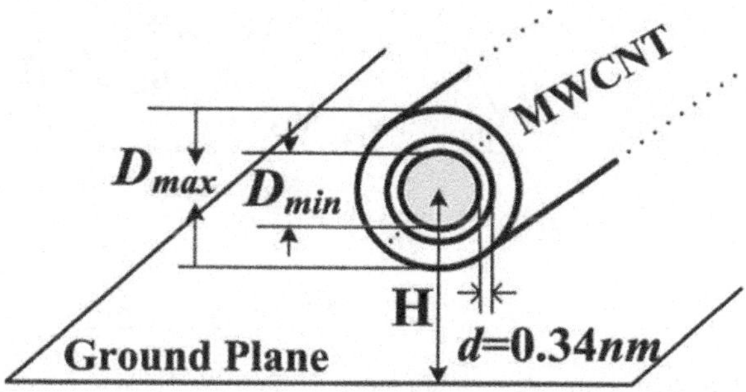

Fig. 2.15 MWCNT placed on a ground plane

$$N_{shells} = 1 + int\frac{D_{max} - D_{max}/2}{2d} \tag{2.17}$$

where int stands for integer part, D_{max} is the diameter of the outer shell and d is the inter-tube distance of 0.34 nm. The resistance of an MWCNT is essentially dependent on the number of conducting channels and can be expressed as

$$R_{MWCNT} = R_Q + R_S.L = \frac{h}{2e^2N}\frac{L}{\lambda} \tag{2.18}$$

where $h/2e^2$ is the intrinsic quantum resistance R_Q (12.9 kΩ), and L, λ and N are the length, mean free path and number of conducting channels of an MWCNT. The kinetic inductance per channel is

$$L_{K/channel} = \frac{h}{2e^2 v_f} \times \frac{1}{2} \approx 8\,\text{nH}/\mu\text{m} \tag{2.19}$$

$$L_{K/shell} = L_{K/channel}/(a.D + b) \tag{2.20}$$

where D is the diamcter of the shell, $a = 0.0612\,\text{nm}^{-1}$ and $b = 0.425$. Therefore, as the diameter increases from 3 to 100 nm, the $L_{K/shell}$ decreases from 13 to 1 nH/μ m. Similarly, the quantum capacitance can be expressed as

$$C_{Q/channel} = 2 \times \frac{2e^2}{hv_f} \approx 193\,\text{aF}/\mu\text{m} \tag{2.21}$$

$$C_{Q/shell} = C_{Q/channel}x(a.D + b) \tag{2.22}$$

Finally, the inter-shell coupling capacitance can be given as

$$C_S = \frac{2\pi\epsilon}{ln(D_{out}/D_{in})} \tag{2.23}$$

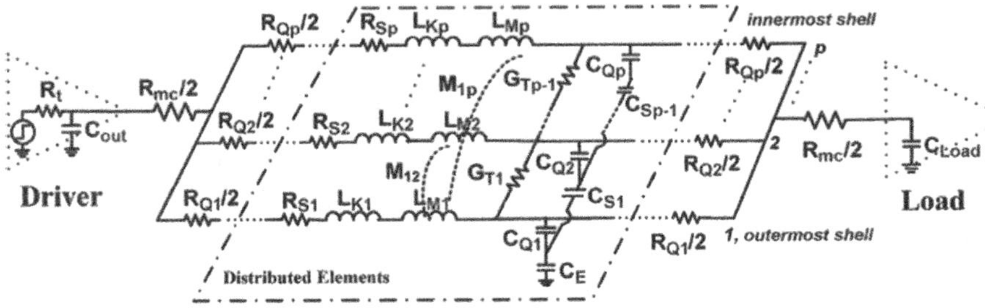

Fig. 2.16 Equivalent circuit model of an MWCNT interconnect (Li et al. 2008)

So, the equivalent TL circuit of a p-shell MWCNT can be drawn as depicted in Fig. 2.16. For MWCNTs, smaller D_{in}/D_{out} indicates more shells in an MWCNT and hence lower resistance. MWCNT structures cannot be converted into simple circuit models since they have shells with varying diameters. Different shells have different channel numbers and varying mean free paths, giving rise to different circuit parameters. The potentials are different for different diameters and hence shell-to-shell capacitance is present. This will be very high due to the small separation between adjacent shells. MWCNTs exhibit the tunnelling phenomenon. Performance analysis of MWCNT interconnects with SWCNT bundle and Cu interconnects by Li et al. (2008) reveals that they are the most suitable interconnects in local, intermediate and global levels. The main reason behind this fact is that MWCNTs have very small electrostatic capacitance compared to SWCNT bundle interconnects. Also, the tube density and the number of conducting CNTs in a SWCNT bundle affect its performance. This aspect is studied in detail in Sect. 2.3.3. The propagation delay of MWCNT interconnects is 15% lesser than Cu wires. In 2008, scientists had experimentally shown that MWCNT interconnects can indeed conduct current at a frequency of 1 GHz (Close et al. 2008). They used MWCNTs as interconnects in 256 ring oscillators that were fabricated by 0.25 μm CMOS technology as shown in Fig. 2.17.

In their experimental setup, an array of gaps was patterned using gold electrodes. Then MWCNTs of around 5 m length and 70 nm diameter were dispersed in ethanol. By applying an AC current, they were able to precisely place the MWCNTs in the gaps using dielectrophoresis effect. The tentative process flow is depicted in Fig. 2.18.

In the first 7 years of the last decade, scientists focussed mainly on SWCNT and MWCNT interconnect applications. Later, however, scientists have proved that realistic nanotube bundles contain both SWCNTs and MWCNTs (Zhu et al. 2006; Li et al. 2003; Haruehanroengra and Wang 2007). A mixed CNT bundle (MCB) is hence a mixture of both SWCNTs and MWCNTs as illustrated in Fig. 2.19. MCBs are considered as large diameter interconnects and possess higher conductance than its counterparts. The impact of tube density, tube distribution, metallic tube ratio, ratio of inner-to-outer tube diameter, bundle width and bundle length on the conductance of the bundle were analysed. Conductance of MCB increases as the bundle width

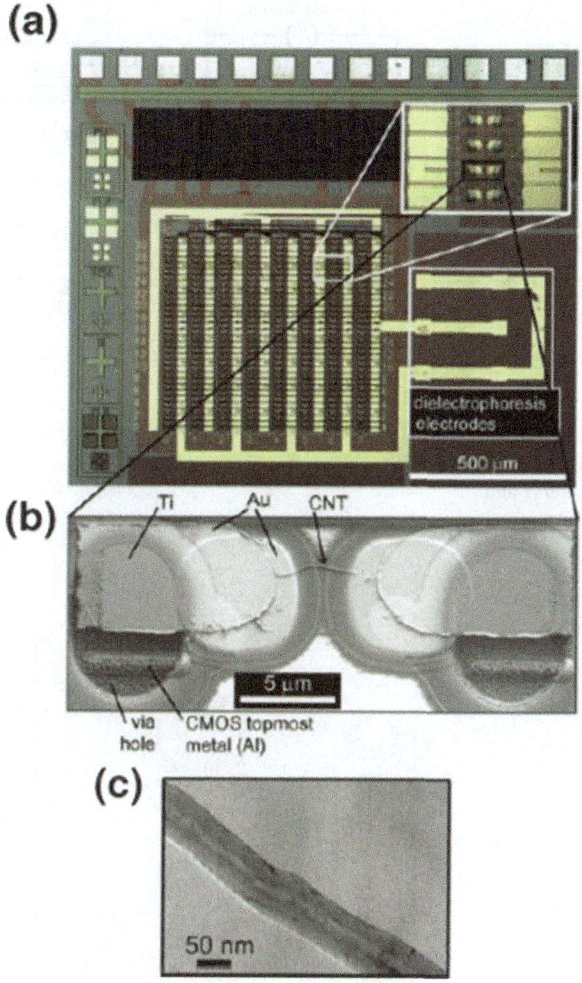

Fig. 2.17 CMOS platform chip assembly of the MWCNT interconnects on the chip surface. **a** Chip photograph. **b** Close up SEM image of one fabricated MWCNT interconnect on top of the chip. **c** TEM image of a typical MWCNT used (Close et al. 2008)

increases (McEuen et al. 2002). Tube densities of 10^{12} tubes/cm^2 have already been achieved. MCBs have at least two-thirds of CNTs as metallic compared to one-third for SWCNT bundles. The tube distribution plays an equal role in determining the conductance of the bundle. Since MWCNTs have a large number of conducting channels, they, along with metallic SWCNTs, should form the core part of the mixed CNT bundle. Semiconducting SWCNTs and small diameter MWCNTs must form the periphery. This will enhance the conductivity and immunity to crosstalk. However, there was a contradiction in this regard, that is, only SWCNTs should be in the core of the bundle rather than MWCNTs (Li et al. 2005). This was due to the misconception

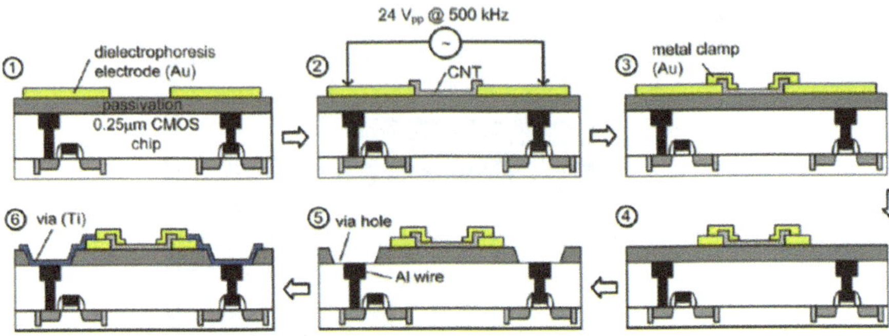

Fig. 2.18 Process flow for fabricating MWCNT interconnects on a CMOS chip (Close et al. 2008)

Fig. 2.19 Mixed CNT bundle that contains SWCNTs of diameter *d* and MWCNTs of various diameters *d* with $D_{min} \leq d \leq D_{max}$ (Haruehanroengra and Wang 2007)

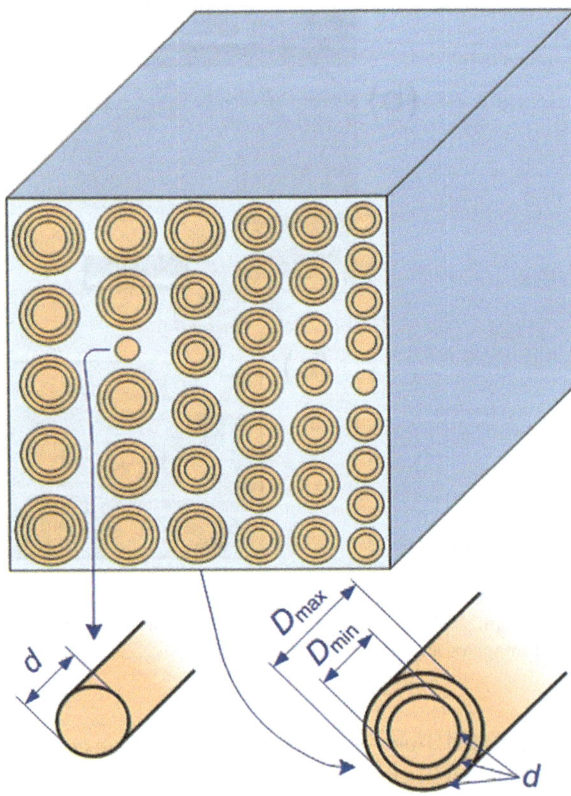

that not all shells in the MWCNTs are conducting. Hence, the role of MWCNTs was considered to be confined as shielding to inner SWCNTs from electromagnetic effects that arise from neighbouring CNT interconnects (Subash and Chowdhury 2009; Kreupl et al. 2002). Mixed CNT bundle conductance was derived from the total number of shells, shell diameter and number of conducting channels per shell. Conductance was derived as (McEuen et al. 2002)

$$G_{bundle} = \sum G_{MW}(D_{outer}, l).N D_{outer} \delta D_{douter} \quad (2.24)$$

where $N(D_{outer})$ is the tube count according to $D'_{outer}s$, and its distribution is a normal (Gaussian) distribution with a mean diameter D_{outer} and a standard deviation σD_{outer}. The conductance analysis of MCBs provides the tube density, tube distribution, metallic tube ratio, inner-to-outer diameter ratio, tube length and tube width. Thus, a bundle of large diameter has more CNTs in it and hence improved conductance. Also, an increase in length beyond mean free path will reduce the conductance. However, MCBs possess very large mean free paths and have better conductance than Cu (Haruehanroengra and Wang 2007). Imperfect contacts between nanotubes and the metal lead to unequal potentials in individual CNTs in a mixed CNT bundle. Thus, electrons tend to flow from one CNT to another depending on the potential of the CNT (Li et al. 2008) implying the presence of tunnelling conductance (G_T) in MCBs. However, as G_T is diameter dependent, tunnelling occurs only from large diameter MWCNTs to SWCNTs. G_T can be expressed as

$$G_T = \sigma.\pi D \quad (2.25)$$

where D is tube diameter and σ is normalized tunnelling conductivity. There are three types of resistances in an interconnect (McEuen et al. 2002). Metal contact resistance that arises due to the imperfect metal nanotube contacts, intrinsic resistance of a CNT that is 6.45 kΩ and scattering resistance due to the phonon scattering in nanotubes. Due to the 1D nature of CNTs, electrons will only be back scattered and hence it determines the mean free path of a nanotube. The magnetic inductance depends on the magnetic field around and inside a bundle. The total kinetic inductance of an MCB depends on the total number of channels and number of shells. For two channels which represent spin up and spin down of electrons, L_K per channel is 8 nH/μ m. Diameter of the tube determines the number of conducting channels. Metallic SWCNTs have N = 2 while semiconducting tubes possess N = 0. MWCNT shells with diameter >4 nm have conducting sub-bands. Also, large diameter semiconducting shells will have some sub-bands as its energy gap is reduced. Thus, large diameter MWCNTs are preferred in MCB interconnects. The overall kinetic inductance is given as

$$L_{K,bundle} = \frac{L_{Kchan}}{\sum_{all Douter's}(N(D_{outer} \times \sum_{Nshell} N_{channel/shell}))} \text{nH}/\mu\text{m} \quad (2.26)$$

where L_{Kchan} is kinetic inductance per conduction channel, $N(D_{outer})$ is the tube count according to Douters, Douter is outer shell diameter of an MWCNT and $N_{channel/shell}$ is the summation of all conducting channels and sub-bands per shell.

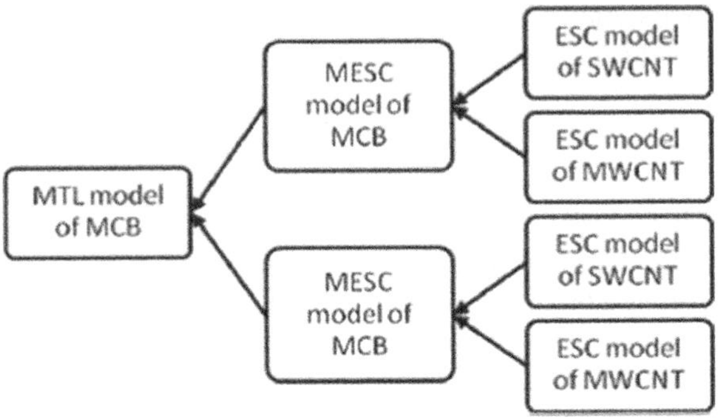

Fig. 2.20 Hierarchical model of mixed CNT bundle interconnects (Sathyakam and Mallick 2011)

Single conductor transmission line modelling of both MWCNTs was carried out by Sarto and Tamburrano (2010). They discussed the equivalent single conductor TL model that is derived from the complex multiconductor TL model. Transient voltage analysis of MESC and MTL models to find delays of SWCNT bundles and MWCNTs was carried out by D'amore et al. (2010). The crosstalk and delay were compared with the 50% time delay of the multiconductor TL (MTL) model. It was pointed out that the MESC model is the simplest approach for finding effects induced by crosstalk coupling. MWCNTs outperform SWCNT bundles in terms of delay. The outer shells of MWCNTs, compared to the inner ones, represent a low-impedance path at high frequencies. Here, an accurate expression for quantum capacitance was derived which includes the effects of mutual electrostatic capacitances, external tube capacitance and shell quantum capacitance.

An accurate and realistic modelling approach based on hierarchical modelling of mixed CNT bundle interconnects was proposed as shown in Fig. 2.20. They carried out MTL as well as MESC modelling for MCB interconnects. The delay values obtained by transient analysis of both the models show that the MCB interconnects are superior to SWCNT bundles as well as MWCNTs. Based on the above observations they proposed the relative positioning of CNTs in an MCB for best performance. They showed that metallic SWCNTs and large diameter MWCNTs should form the core of the bundle as shown in Fig. 2.21.

Also, semiconducting SWCNTs and small diameter MWCNTs must be at the periphery so that signals flow through the core and the wire is isolated from crosstalk. More recently, many works were carried out to analyse various types of mixed CNT bundle interconnects. First, Majumder et al. (2012b) had analysed the dynamic crosstalk-based analysis of mixed CNT bundles. They proposed four configurations of MCBs as shown in Fig. 2.22.

The crosstalk-induced delay was studied at various interconnect lengths ranging from 100 to 1000 μm. It was found that the crosstalk-induced delay was least for case 1 where MWCNTs are at the periphery and SWCNTs are at the core. Later in 2014, Majumder, Kaushik and Manhas (2014) had compared the performance of MCB interconnects of the types shown in Figs. 2.22 and 2.23 with that of SWCNT bundle and MWCNT bundle interconnects. They incorporated the effects of coupling

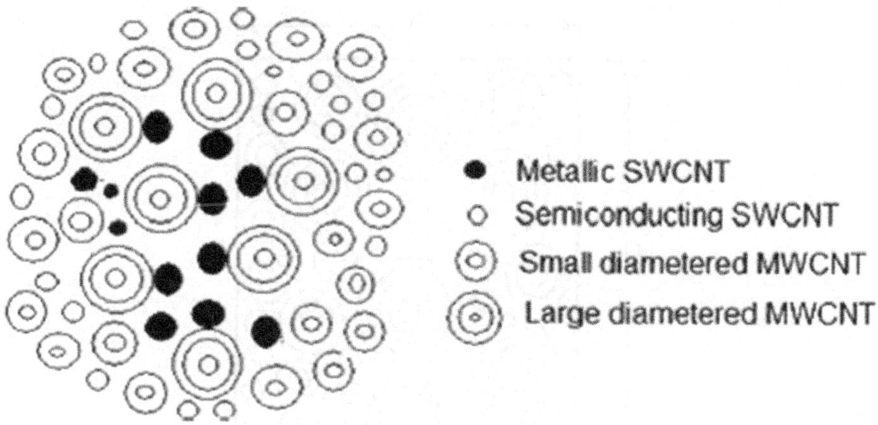

Fig. 2.21 Proposed MCB geometry (Sathyakam and Mallick 2012)

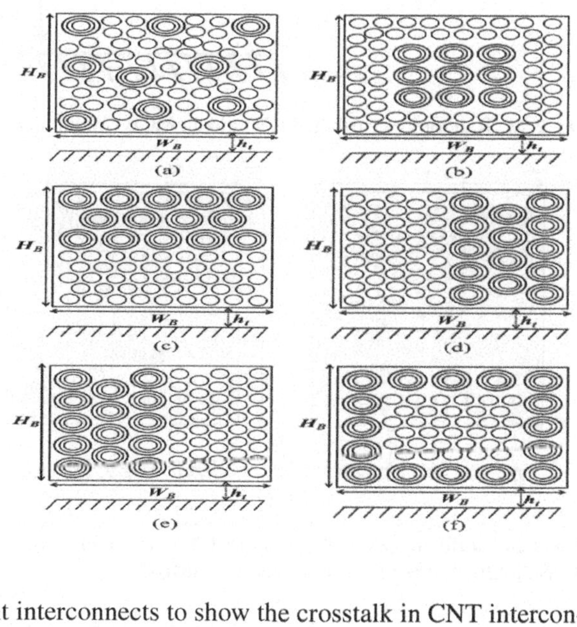

Fig. 2.22 Various types of MCBs: **a** MCB-1, **b** MCB-2, **c** MCB-3, **d** MCB-4 (Majumder, Kaushik and Manhas 2014)

capacitance between adjacent interconnects to show the crosstalk in CNT interconnects. SPICE-based simulations of the ESC models show that the RC delay and crosstalk delay are significantly smaller for MCB-2 where MWCNTs are at the core and SWCNTs are at the periphery.

Similarly, the performance of a randomly distributed MCB interconnect is analysed by Majumder et al. (2015) as shown in Fig. 2.23. They varied the count of SWCNTs and MWCNTs in the bundle. For various configurations, crosstalk analysis was performed.

Results showed that MWCNT bundles with more number of shells have less delay than the bundle that has SWCNTs and MWCNTs of smaller number of shells. More

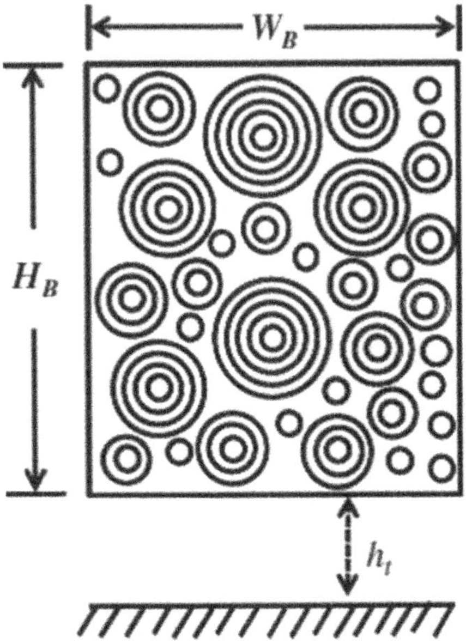

Fig. 2.23 Randomly distributed CNTs in an MCB (Majumder, Das and Kaushik 2014)

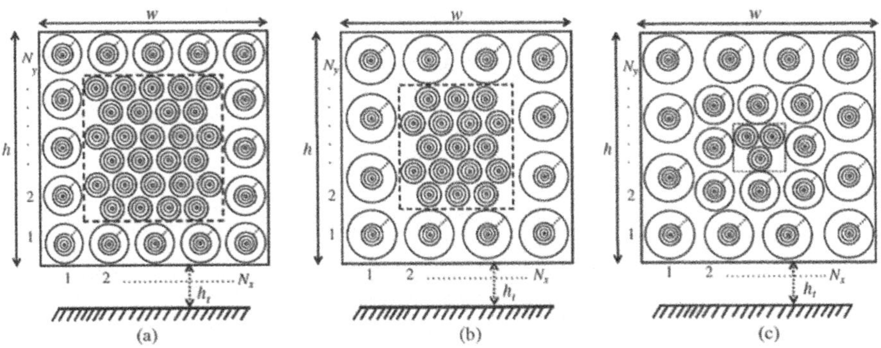

Fig. 2.24 Different types of mixed MWCNT bundle interconnects studied by Majumder, Das and Kaushik (2014) **a** MMB-1, **b** MMB-2, **c** MMB-3

recently Majumder, Das and Kaushik (2014) worked on delay analysis of mixed MWCNT bundle interconnects which are shown in Fig. 2.24. This is a remarkable work as the MCBs considered are more realistic than the ones proposed earlier. Each mixed MWCNT bundle (MMB) interconnect has large diameter MWCNTs placed at the periphery and small diameter MWCNTs placed at the core. The number of small diameter MWCNTs placed at the core varies for each case of MMB as shown in Fig. 2.24. Results show that the MMB-3 structure performs best in terms of propagation delay and crosstalk delay compared to conventionally used MWCNT bundle interconnects of the same dimensions. The propagation delay is 15% lesser and the crosstalk delay is 29% lesser.

2.3.2 *Design Metrics of CNT Interconnects*

In this section, various design metrics that are important in improving the performance of CNT interconnects are discussed. The main factors are

- number of conducting channels,
- nanotube count in a bundle and
- nanotube density in a bundle.

These factors are discussed in detail in the following sections.

2.3.3 *Number of Conducting Channels in a CNT Bundle*

The importance of using bundles of SWCNTs, MWCNTs or a mix of both SW and MWCNTs as interconnects was seen in the preceding sections. Basically, by using CNTs as bundled interconnects rather than a single CNT, the total number of conducting channels that contribute to electrical conduction in an interconnect increases. Hence, the current-carrying capacity of the interconnect increases. The number of channels available for conduction is dependent on the CNT diameter D and the temperature of the surrounding medium T. Furthermore, the number of conducting channels is only two-third of the total number of channels when the CNT diameter is less than d_T/T where $d_T = 1300$ nmK. Thus, for high-frequency conduction of electrons at long mean free paths, the number of conducting channels should be as high as possible. For that, large diameter CNTs are preferred in CNT bundle interconnects. The number of conducting channels increases with temperature as can be seen from Eq. 2.16. Figure 2.25a shows the number of conduction channels versus MWCNT diameter.

2.3.4 *Dimensions and Tube Count of a CNT Bundle*

The number of metallic carbon nanotubes in a CNT bundle is very important as it determines the overall conductivity which directly affects the performance. The more the number of CNTs in a CNT bundle, the higher its conductivity. But the intrinsic quantum capacitance will increase too. Therefore, the two main factors that determine the number of CNTs in a bundle are the RC time constant of the bundle and the technology node dimensions that are stipulated by the ITRS (2013). Table 2.2 shows the nominal dimensions of interconnects at each technology node.

Table 2.2 Nominal interconnect dimensions as per ITRS 2013 recommendations

Technology node	Width (nm)	Height (nm)	Aspect ratio (H/W)
18 nm	45	90	2.0
14 nm	32	67	2.1
12 nm	25	52	2.1
10 nm	20	44	2.2
7 nm	14	31	2.2

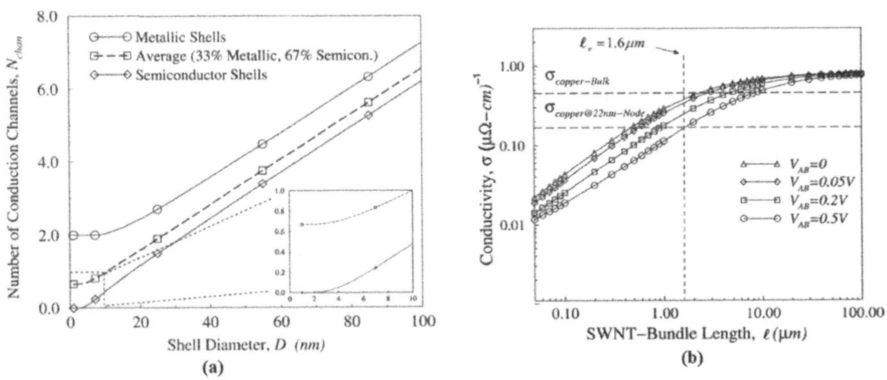

Fig. 2.25 **a** Number of conducting channels versus shell diameter (Naeemi and Meindl 2006), **b** Conductivity versus SWCNT bundle length (Naeemi and Meindl 2007b)

Based on the technology node width, CNT interconnects are modelled as square bundles where the width and height are the same and the aspect ratio (AR) is unity. This is unlike in Cu wires whose AR is considered as 1.5 or 2. The expressions for finding the number of SWCNTs for any given dimensions of width and height are given in Eq. 2.9. So, the width and the height of a CNT interconnect are very important in determining its resistance and capacitance. Furthermore, its performance is highly dependent on its length.

As shown by Naeemi and Meindl (2007b) in Fig. 2.25, SWCNT bundle conductivity increases with wire length up to 10 μm. This is due to the reduced scattering of electrons down to 1.6 μm, and subsequently the conductivity saturates at a point higher than that of bulk Cu for all bias voltages from 0 to 0.5V. This is precisely the reason behind the preference of CNTs over Cu as interconnects. The latency or delay of the interconnect is dependent on the interconnect length. The effect of wire width on its performance is profound. From Fig. 2.26, it can be noted that the speed enhancement of the signals in the wire reduces with increasing the wire width. In terms of the delay, as the width increases, delay increases due to high wire to ground electrostatic capacitance. However, keeping in mind the aspect ratio, which increases as the width increases, the energy per switching decreases as the width and hence the aspect ratio increases as shown in Fig. 2.27. The overall resistance of the CNT

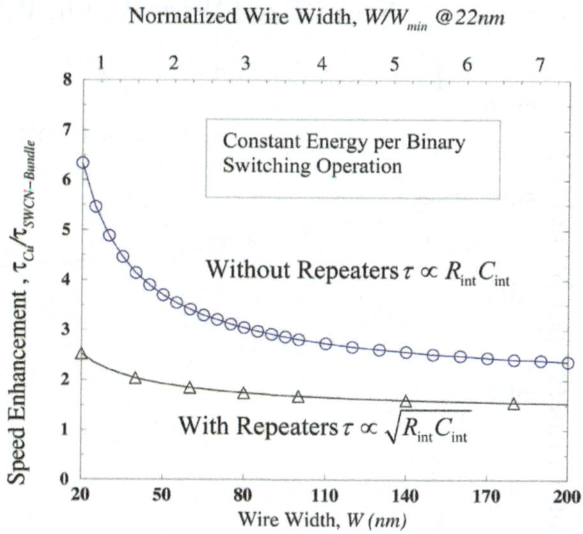

Fig. 2.26 Speed enhancement versus wire width (Naeemi and Meindl 2007b)

bundle decreases as the bundle width increases and the bundle length decreases. The individual SWCNTs have a diameter of 1.0 nm here. Thus, there must be an optimum width and length, when the interconnect resistance does not exceed that of Cu wires as per Nieuwoudt and Massoud (2006).

Consequently, it is inevitable to decrease the wire width as the technology node size decreases every 2 years as depicted in Fig. 2.28.

2.3.5 *Carbon Nanotube Density in a CNT Bundle*

Carbon nanotubes can be used as interconnects in either bundle form or rope form depending on the fabrication methods. The density of the CNTs in the bundle affects its operation and performance. Ideally, a CNT bundle must have all CNTs metallic in it for highest possible conduction. In practice, however, due to the non-idealities during fabrication, a bundle can have up to 80% CNTs as metallic (Alizadeh and Sarvari 2016). It can be expected that, as the fabrication processes mature, it is possible to have almost all CNTs as metallic in a bundle. Figure 2.29 shows the capacitance and resistance changes due to the change in metallic CNT (mCNT) density in the SWCNT bundle (Srivastava et al. 2009). More specifically, the capacitance increases within an order of magnitude for all cases including the ones shown in Fig. 2.30. Particularly, the capacitance is smaller for 14 nm technology node size than at 22 nm node size. Additionally, the electrostatic and the quantum capacitances follow the same trend with respect to mCNT density.

Finally, the delay of the SWCNT bundle was found to be affected by the metallic tube density in the bundle. It was found that for 80% of mCNTs in the bundle, the

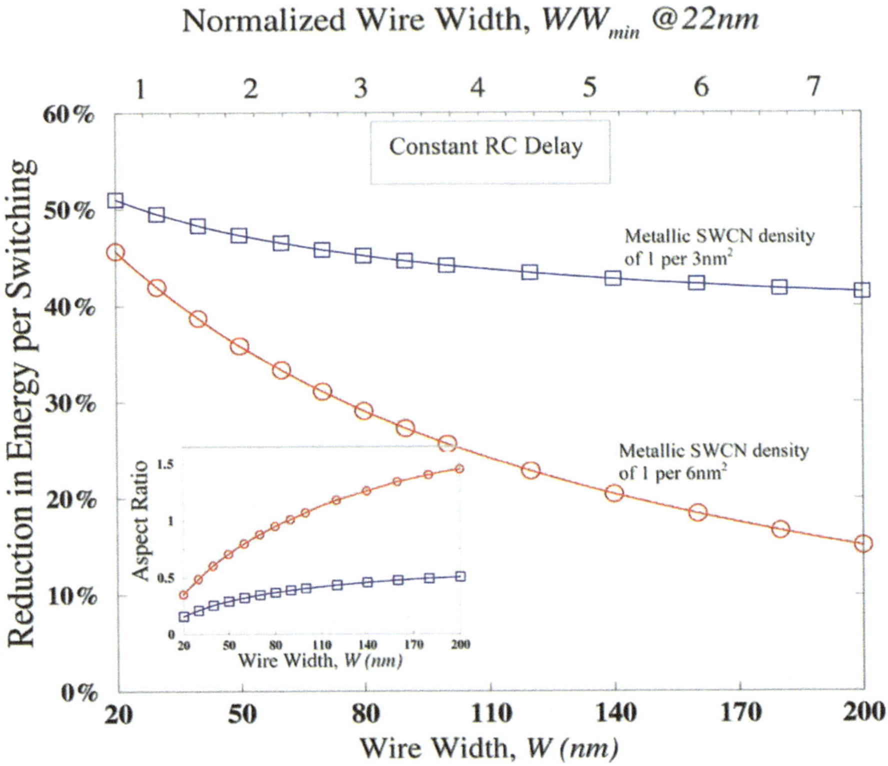

Fig. 2.27 Reduction in energy per switching versus wire width (Nieuwoudt and Massoud 2006)

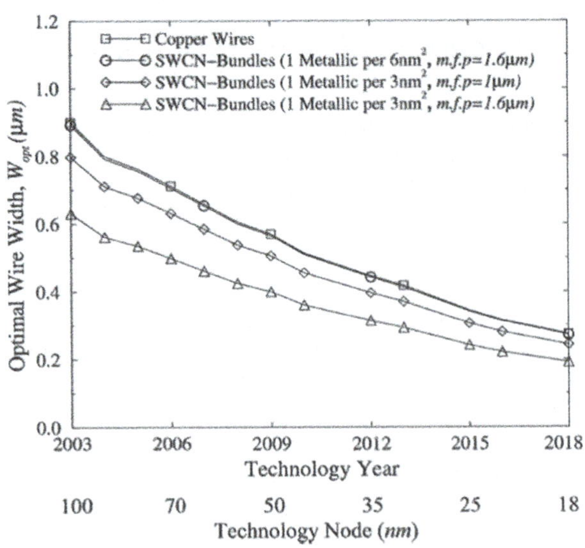

Fig. 2.28 Optimal wire width versus technology node (Nieuwoudt and Massoud 2006)

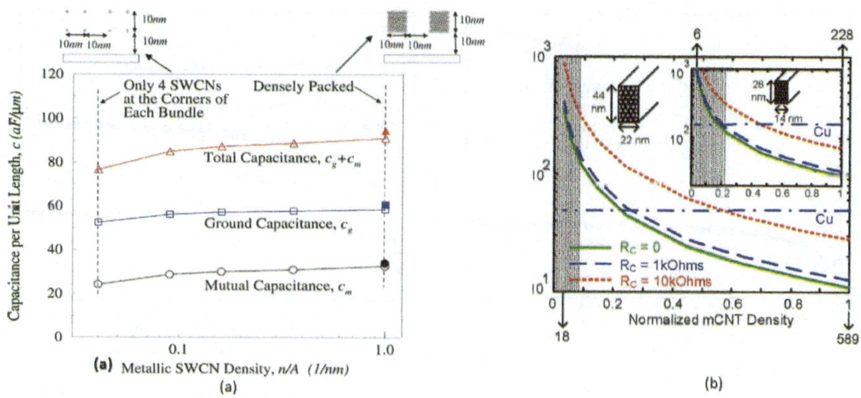

Fig. 2.29 **a** Capacitance and **b** Resistance versus SWCNT bundle density (Srivastava et al. 2009)

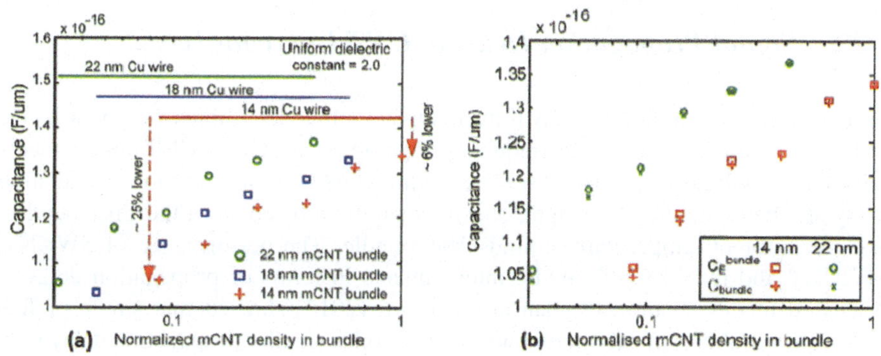

Fig. 2.30 **a** Capacitance at different technology nodes, **b** Electrostatic and bundle capacitance versus normalized CNT bundle density (Srivastava et al. 2009)

Fig. 2.31 Bundled SWCNT (BSWNT) delay versus temperature with varying mCNT density (Alizadeh and Sarvari 2016)

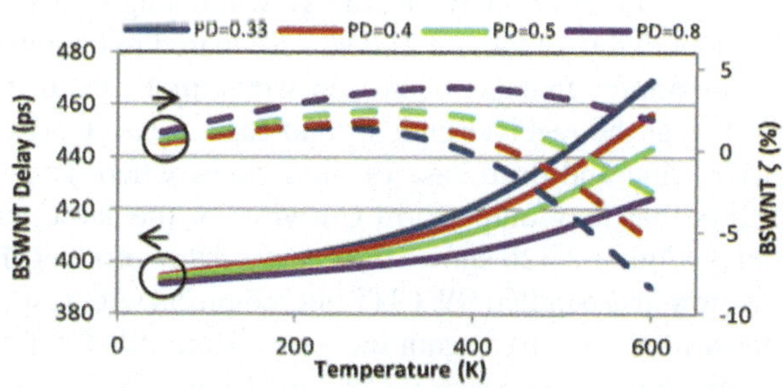

delay is least compared to 33%, 40% and 50% of mCNTs in the bundle as shown in Fig. 2.31 (Alizadeh and Sarvari 2016).

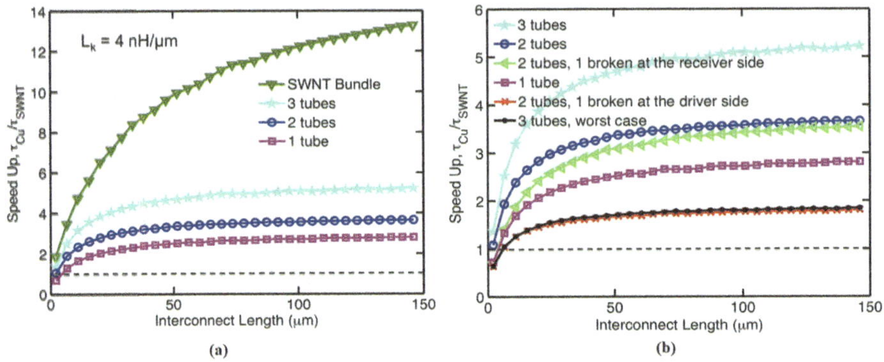

Fig. 2.32 Comparison of **a** propagation delay enhancement and **b** energy-delay product of CNTs w.r.t Cu (Ceyhan and Naeemi 2013)

2.3.6 Signal Propagation Delay in CNT Interconnects

Propagation delay or the RC delay is the main factor that determines the performance of a CNT interconnect. For high-frequency operation ($f > 1$ GHz), CNT interconnects must have small propagation delay. The main factors that influence the propagation delay are the width, height, length, density, metallic tube count of the CNT bundles and the ambient temperature around CNT bundle. The performance of SWCNT, MWCNT and mixed CNT bundle interconnects in terms of propagation delay is discussed in this section. Ceyhan and Naeemi (2013) showed that single or few CNTs are suitable as local interconnects due to their smaller propagation delay than Cu wires.

Figure 2.32a shows the propagation delay enhancement of single and few CNT interconnects with respect to Cu interconnects. The delay analysis reveals that SWCNT bundles have least delay while single, two and three CNT interconnects have moderate reduction in delay compared to Cu interconnects of the same length. Furthermore, analysis of broken wires, first case of two CNTs where one CNT is broken at the receiver end, second case of two CNTs where one CNT is broken at driver side and third case of three tubes where two are broken at the driver side, shows that they outperform Cu wires of the same length in terms of propagation delay. Figure 2.32b shows the energy-delay product (EDP) gain comparison of single, few and bundled SWCNT interconnects with respect to Cu interconnects. As the delay increases, EDP gain increases. Here, SWCNT bundles have larger EDP gain compared to other cases of CNT and Cu interconnects at 150 μm length. When some CNTs are broken in a bundle, as seen previously, the EDP gain of three CNT interconnects is highest among other cases of CNT interconnects. In the case of SWCNT bundle interconnects, the delay ratio of CNT over Cu interconnects is plotted in Fig. 2.33. In the case of intermediate interconnects whose length is up to 500 μm, the delay of CNT interconnects is far lesser than Cu wires at 14 nm technology node (Fig. 2.34a). In the case of global interconnects, the delay ratio of CNT over Cu shows

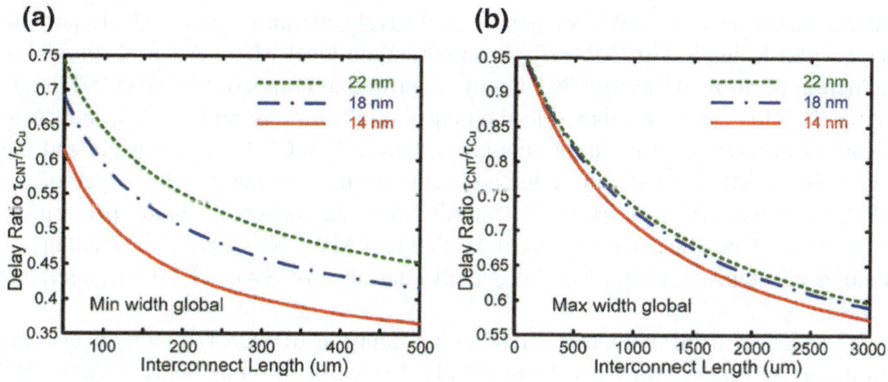

Fig. 2.33 Delay enhancement for **a** intermediate and **b** global interconnect lengths at different technology nodes (Ceyhan and Naeemi 2013)

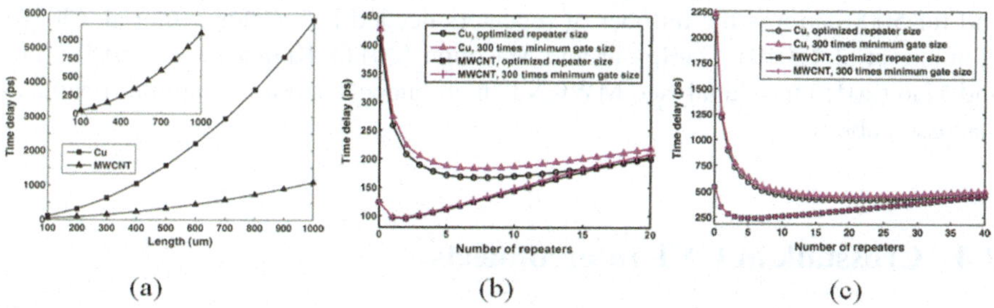

Fig. 2.34 **a** Delay versus length **b** delay versus number of repeaters for 2000-m-long wires and **c** delay versus number of repeaters for 5000-m-long wires (Li et al. 2008)

that the delay ratio of CNT interconnects over Cu interconnects reduces drastically from 1 μm to 3000 μm length (Figs. 2.34b).

MWCNTs have received equal attention as SWCNTs in the analysis of propagation delay and in comparison with Cu interconnects. Li et al. (2008) had first analysed the propagation delay of MWCNT interconnects. Figure 2.35 shows the improvement in delay of MWCNT wires over Cu wires at different interconnect lengths from 200 to 1200 μm. In all cases, the MWCNT interconnect delay is lesser than Cu interconnect delay. Furthermore, from Fig. 2.35a, it can be observed that for an AR of 3, the delay decreases with wire width. The delay is higher when the inter-shell tunnelling conductance is considered compared to the case when it is not considered. In Fig. 2.35b, the delay increases with line width for 22 nm node, constant height and MWCNT diameter. In Fig. 2.35c, the delay decreases when the MWCNT diameter increases.

Comparison of delay of SWCNT bundle and MWCNT interconnects shows that, MWCNT interconnects perform better than SWCNT bundle interconnects when the SWCNT bundle involves random chirality CNTs in it (Li et al. 2008). In particular, SWCNT bundle interconnects perform in the same way or even better than MWCNT interconnects when the SWCNT bundle exclusively contains metallic CNTs and the wire width is small with AR = 3. At intermediate level where AR = 2, the delay reduction of all metallic SWCNT bundle is profound compared to MWCNT interconnects. This can be attributed to the contribution of more number of conducting channels in SWCNT bundle interconnects than in MWCNT interconnects and the reduction in AR coupled with reduction in the intrinsic quantum capacitance of the SWCNT bundle. Also, at local level, MWCNT interconnects are better performers than SWCNT bundles. This is due to the fact that MWCNTs have the advantage of smaller resistance compared to the overall resistance of SWCNT bundle interconnects.

The idea of using repeaters for reducing delay in MWCNT interconnects was proposed by Liang, Wang and Ding (2011). Basically, the time delay of MWCNT interconnects was found to be smaller than Cu interconnects up to 1000 μm. However, to keep up with this trend for higher lengths up to 5000 μm, repeaters are necessary. It was found that for lengths up to 2000 μm, two repeaters can optimize the delay and for 5000 μm and the number of repeaters needed is 5. Later, Xu et al. (2010), D'amore et al. (2010), Gholipour and Masoumi (2011), Kumar et al. (2015), Tang and Mao (2015) have analysed MWCNT interconnects and found similar results as discussed above.

2.4 Crosstalk in CNT Interconnects

2.4.1 *Status and Challenges*

Crosstalk is one of the foremost important consequences of down scaling of components in integrated circuits. The major factors that influence the crosstalk between two adjacent interconnects are

- Distance between adjacent interconnects,
- Aspect ratio (H/W) of interconnects,
- Dielectric medium between adjacent interconnects and
- Switching signal (either 0 or 1).

As the distance between adjacent interconnects decreases, the electrostatic coupling between the interconnects increases which leads to high coupling capacitance. The resulting coupling capacitance will be highest when both the interconnects are simultaneously at logic level 1 at the same time. Coupling capacitance leads to crosstalk between interconnects. The crosstalk is worst when the interconnects are switching opposite signals of 0 or 1.

Figure 2.35a shows the schematic of two capacitively coupled interconnects. The dielectric medium between the interconnects is a dielectric material that has low dielectric constant. As the dielectric constant decreases, coupling capacitance and hence crosstalk can be reduced. Ideally, if air is used as the medium, the dielectric

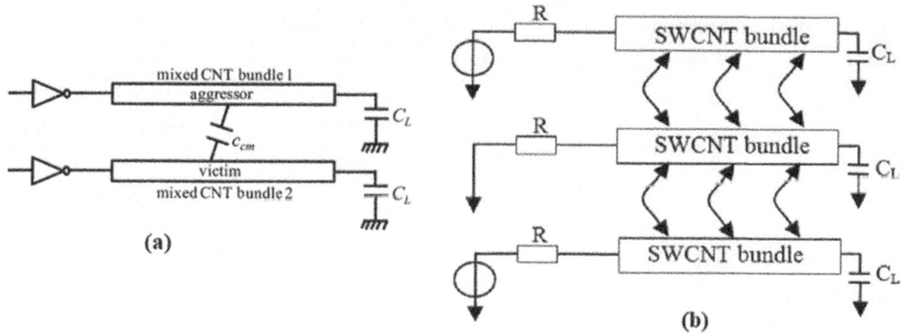

Fig. 2.35 **a** Capacitively coupled mixed CNT bundle interconnects, **b** Three coupled interconnects

constant is unity which can significantly reduce the coupling capacitance. The height of the interconnect influences the coupling capacitance. If the height is reduced until it equals the width, the aspect ratio becomes unity, and consequently the coupling capacitance reduces. Crosstalk in interconnects was extensively studied by Kaushik et al. (2007), Kaushik and Sarkar (2008a, b). They have developed an aggressor–victim model of interconnects where the energized line is the aggressor and the idle line is the victim. Due to the signal transmission through the aggressor line, an e.m.f. is induced on the victim line whenever the aggressor line passes a logic 1. The resulting induced e.m.f is called as the crosstalk and the voltage magnitude is called as the peak noise. Crosstalk effects are extensively studied in CNT interconnects. In the following sections, the status of research on crosstalk in CNT interconnects is covered.

Crosstalk was found to be a major challenge in CNT bundle interconnects. Single and few CNT interconnects and monolayer interconnects, which are short in length, will not have crosstalk effects. Bundled SWCNT, MWCNT and mixed CNT bundle interconnects, which are used as long intermediate/global interconnects, can be significantly affected by crosstalk. Many researchers have analysed crosstalk in CNT interconnects. For example, Rossi et al. (2007), Pu et al. (2009), Nieuwoudt and Massoud (2006), Das and Rahaman (2011), and Majumder et al. (2012b) had studied crosstalk effects in CNT interconnects. Initial works until 2012 were focussed on crosstalk analysis of SWCNT bundle interconnects. Later works until present discuss the crosstalk in MWCNT interconnects and mixed CNT bundle interconnects.

2.4.2 Aggressor–Victim Model

CNT interconnects are driven by CMOS inverters as discussed in previous sections. In densely packed ICs, adjacent CNT interconnects are separated by a few tens of nanometres. This distance is usually considered as the same as width (W) of the

Table 2.3 Various switching conditions in two adjacent interconnect lines

Line-1	Line-2	Crosstalk	Noise	Remarks
0	0	Nil	Nil	Both lines are idle
0	1	High	High	Line-1 is victim
1	0	High	High	Line-2 is Victim
1	1	No effects	High	Both lines are energized

interconnect. Figure 2.35b shows two adjacently placed CNT bundle interconnects on a substrate.

One of the interconnects is the aggressor and the other, the victim or vice versa. The coupling capacitance of the arrangement shown in Fig. 2.35b can be given as

$$C_C = \frac{2\pi\epsilon}{cosh^{-1}(y/d)} \tag{2.27}$$

Table 2.3 shows the possible states of switching conditions of two adjacent lines.

The crosstalk is high when the two lines are switching in the opposite levels. Miller capacitance effects are considered in previous works where crosstalk in CNT interconnects is modelled (Majumder, Kaushik and Manhas 2014) as depicted in Table 2.4. The work in this book considers the effects of Miller capacitances while computing crosstalk effects in CNT interconnects.

Sapatnekar (2004) gave a detailed overview and analysis of timing in VLSI signals. Miller capacitance is discussed where three conditions are possible. The first case is when one line is switching and the other one is idle. Here, the Miller capacitance to the ground is equal to the coupling capacitance between the two lines. The second case is considered when the aggressor and the victim are switching in the same direction, in which the Miller capacitance is equal to zero. Last case is considered when the lines are switching in the opposite directions, in which the Miller capacitance is double the coupling capacitance.

Crosstalk is a major performance bottleneck in VLSI circuits. Das and Rahaman (2011) showed that crosstalk can affect the system timing, functionality and reliability of circuits depending on the type of switching activity between the aggressor and victim lines as shown in Table 2.4.

2.4.3 Crosstalk in Three Coupled Interconnects

A more realistic case is when three switching lines are adjacent to each other as shown in Fig. 2.36.

Crosstalk analysis was done by Zhang et al. (2012) for 100, 200 and 300 μm lengths of interconnects as shown in Fig. 2.36a. The results show that due to crosstalk,

Table 2.4 Effects of crosstalk on circuit performance

Aggressor transition	Victim transition	Results	Impacts
Logic 0 to logic 1	Logic 0 to logic 1	Decrease in rise time	Timing
Logic 0 to logic 1	Logic 1 to logic 0	Increase in fall time	Timing
Logic 1 to logic 0	Logic 0 to logic 1	Increase in rise time	Timing
Logic 1 to logic 0	Logic 1 to logic 0	Decrease in fall time	Timing
Logic 0 to logic 1	Held at Logic 1	Overshoot	Reliability
Logic 0 to logic 1	Held at Logic 0	Rise glitch	Functionality
Logic 1 to logic 0	Held at logic 1	Fall glitch	Functionality
Logic 1 to logic 0	Held at logic 0	Undershoot	Reliability

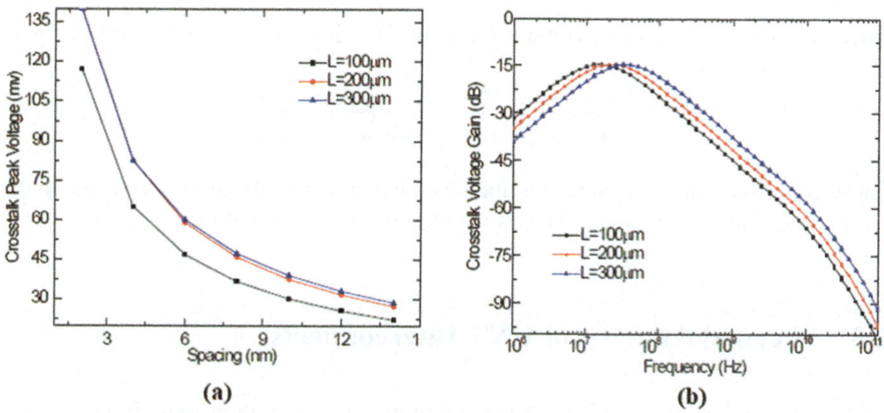

Fig. 2.36 **a** Crosstalk-induced peak noise in SWCNT bundle interconnects for various spacing between the bundles, **b** crosstalk voltage gain versus frequency (Zhang et al. 2012)

a peak noise voltage is induced on the victim lines and it decreases as the line spacing increases, as shown in Fig. 2.36b. The coupling capacitance between the lines decreases as the line spacing increases. More importantly, the crosstalk effects start to diminish at very high frequencies starting from 10^8 Hz as the crosstalk voltage gain reduces rapidly with increasing frequency as shown in Fig. 2.36. This implies that at high-frequency operation, the signal amplitude gets reduced due to crosstalk effects.

2.4.4 *Crosstalk-Induced Peak Noise*

Crosstalk induces peak noise in interconnects. Due to the coupling of the electrostatic fields among the lines they can experience crosstalk overshoot or undershoot of signals which results in noise.

For a two line system, where one line is energized and another line is idle, the crosstalk-induced noise can be given as

$$\Delta V_{victim} = \frac{C_{adj}}{C_{adj} + C_{gnd-v}} \Delta V_{Aggressor} \tag{2.28}$$

where $\Delta V_{Aggressor}$ and ΔV_{Victim} are the input and output voltages at the aggressor and victim, respectively, C_{adj} and C_{gnd-v} are the coupling and electrostatic capacitances, respectively. When both lines are energized, the crosstalk-induced peak noise can be given as

$$\Delta V_{victim} = \frac{C_{adj}}{C_{adj} + C_{gnd-v}} \frac{1}{1+k} \Delta V_{Aggressor} \tag{2.29}$$

where k is the ratio of the aggressor delay time to victim delay time and is given as

$$k = \frac{\tau_{Aggressor}}{\tau_{Victim}} = \frac{R_{Aggressor}(C_{adj} + C_{gnd-a})}{R_{Victim}(C_{adj} + C_{gnd-v})} \tag{2.30}$$

These equations can be used to calculate the peak noise by the aggressor lines on the victim lines and this concept can be extended to three coupled lines as well.

2.5 Thermal Analysis of CNT Interconnects

Carbon nanotubes are excellent thermal conductors along their axis. Experimental results show that CNTs have thermal conductivity of 3500 W/mK (Pop et al. 2006). However, they conduct very little along their circumference. The carrier transport in CNTs is temperature dependent and is influenced by the ambient temperature. Understanding of carrier behaviour in CNTs at various temperatures gives information on its performance as VLSI interconnects. As the operating temperatures of ICs is well above the room temperature, it is desirable to have interconnects that operate satisfactorily at such temperatures. Copper, as it approaches to dimensions at which its resistivity becomes too high to manage, is posing challenges in carrier transport due to its poor thermal conductivity of 400 W/mK (Quick et al. 1895). Thus, carbon nanotubes can solve this problem by conducting both heat and electricity at the same time along their length. The underlying physical mechanisms of electron–phonon and phonon transport are well studied by Naeemi and Meindl (2007c), Chen et al. (2009), Mohsin et al. (2013), Mohsin et al. (2014) and Naeemi and Meindl (2005a). The governing equations for the acoustic and optical phonon mean free path are modelled and these calculations are incorporated in studying the temperature-dependent resistance of CNTs (Kane et al. 1998). The resulting equations are discussed and used for temperature dependence of performance of CNT interconnects in Chap. 7 of this book. Other groups have studied the Joule heating effects in CNTs where the one-dimensional Joule heating equation is applied to CNTs with known current

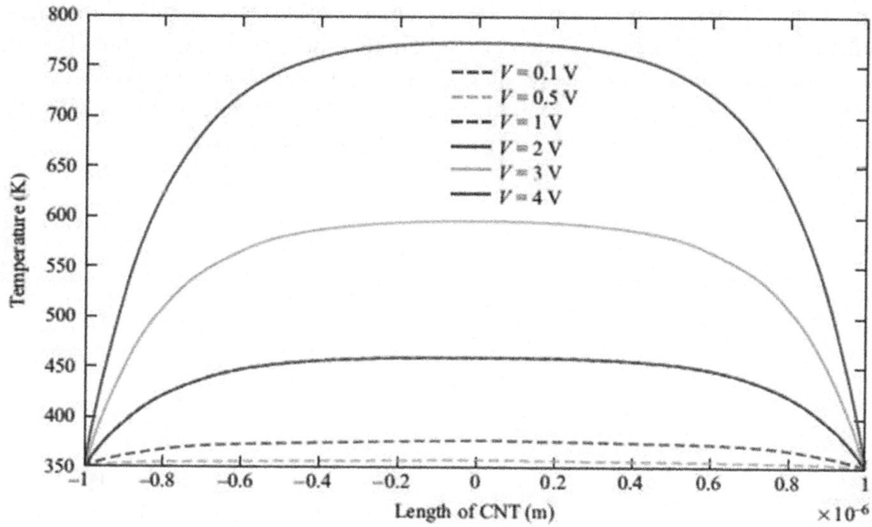

Fig. 2.37 Temperature profile along the length of an SWCNT interconnect (Naeemi and Meindl 2007b)

densities at different temperatures (Naeemi and Meindl 2007c; Chen et al. 2009; and Verma et al. 2011). Solutions show that the temperature is concentrated at the centre of the wire and heat tends to dissipate at the end contacts as depicted in Fig. 2.37. Another study shows the effects of using CNTs on a Si substrate or by suspending them between two contacts (Yamada et al. 2009). It was found that suspended CNTs can suffer from thermal breakage at high temperatures while CNTs placed on Si substrate dissipated heat to the substrate and are less prone to thermal breakage. Later, interconnect modelling based on temperature dependence was done by many researchers including Liang, Wang and Lin (2011) and Singh and Raj (2015a). In this book, the concept of temperature coefficient of resistance discussed by Naeemi and Meindl (2007c) is extended to CNT interconnects with air gaps. Also, the acoustic and optical phonon mean free path and temperature-dependent resistance are considered for the analysis of CNT interconnects under subthreshold conditions.

2.6 Summary

This chapter discussed the background and the status of research in CNT interconnects. Electron transport and Luttinger liquid theory models of carbon nanotubes that are further used for constructing interconnect models are explained. Then, the modelling and simulations carried for SWCNT, MWCNT and mixed CNT bundle interconnects were discussed. Design metrics that affect the performance of CNT interconnects such as number of conducting channels in a CNT bundle, carbon nanotube count in a CNT bundle and carbon nanotube density in a CNT bundle were thoroughly explained. Next, the problems in signal propagation in interconnects, i.e. the RC delay, were discussed for all the three types of CNT interconnects. Crosstalk-induced delay and peak noise in CNT interconnects are discussed.

Chapter 3
Reducing Crosstalk in CNT Interconnects

Crosstalk reduction in CNT interconnects is a major challenge. Some of the ways to reduce crosstalk and induced delay in various CNT bundle structures are discussed in this chapter. First, use of semiconducting CNTs around metallic CNTs is proposed. Surprisingly, semiconducting CNTs are found to have small radial dielectric constants which are useful for shielding or as low-k dielectrics in interconnects. Further, crosstalk in mixed CNT bundles is also discussed and semiconducting CNTs are used to shield mixed CNT bundles. Then, a new CNT interconnect structure, namely, triangular CNT (T-CNT) bundle interconnect is proposed. Due to their inherent geometric advantage, they are least capacitively coupled and offer least possible crosstalk. Results are interesting and T-CNT bundle interconnects can become the ultimate choice in VLSI.

3.1 Background

Interconnect performance in VLSI circuits has become a major challenge now for integrated circuit (IC) designers. Researchers emphasized on the use of carbon nanotubes (CNTs) as a prospective substitute to copper wires in the ICs in the last two decades (Turkane and Kureshi 2017). Recent works show promising results in precisely growing, aligning and testing of various types of CNT nanointerconnects (Wejinya et al. 2013; Zhou et al. 2012; Chai et al. 2011). Unlike Cu technology where trenches are made on Si substrates to deposit molten copper, CNTs were grown on the substrate first and then modified according to the needs. Hence, it is a bottom-up approach rather than the traditional top-down. CNTs are a promising replacement of Cu as they are better in terms of performance and reliability (Ceyhan and Naeemi 2013). However, in scaled technologies, the overall latency in an IC is more dependent on the interconnect capacitance rather than the transistor switch-

P. U. Sathyakam and P. S. Mallick, *Design and Crosstalk Analysis in Carbon Nanotube Interconnects*, https://doi.org/10.1007/978-981-15-8888-4_3

ing period. Reduction of EMI, crosstalk, delay and the effective wire capacitance is of paramount importance in future on-chip nanointerconnects. It was assumed earlier that semiconducting SWCNTs do not contribute to conduction and hence were neglected (Srivastava et al. 2009; Giustiniani et al. 2011). Subash et al. (2013) considered MWCNTs and large diameter SWCNTs as crosstalk shields, which led to high losses in interconnects in terms of delay and performance. Consequently, a perfect EMI-blocking material is still a dream.

3.2 Crosstalk Reduction Using S-CNTs

3.2.1 Role of S-CNTs

In this section, the advantage of using semiconducting SWCNTs (s-CNTS) for future CNT-based interconnects with enhanced performance and reliability is explained. The role of s-CNTs as EMI-reducing material in CNT interconnects is explained for the first time in this book. It is also demonstrated here, how the electrostatic capacitance between the substrate and interconnect, and coupling capacitance between adjacent lines can be reduced. The schematic arrangement of the CNT interconnect structure with nominal dimensions is shown in Fig. 3.1. Semiconducting CNTs are p-type in nature (Rochefort et al. 2001) and hence they carry larger number of holes when energized. They are considered as a prospective material for EMI shielding in CNT interconnects. A thick layer of s-CNTs upon CNT interconnects inhibits the capacitive coupling with respect to its neighbours. The variations in capacitance as a function of s-CNT layer thickness are studied here. Growth of s-CNTs with controlled diameter was already reported by Chai et al. (2011). Only semiconducting zigzag CNTs are considered here. Other semi-metallic or chiral CNTs are not considered because of their small bandgap. In metallic SWCNTs and MWCNTs, the radial conductance is due to the tunnelling of electrons through overlapping π-orbitals. The tunnelling conductance is proportional to the tube diameter, and hence is negligible in SWCNT bundles (Sato et al. 2006). However, radial geometry of zigzag s-CNTs resembles that of metallic graphene. Thus, an array of s-CNTs will have more surface area than a single CNT. Thus, in the present case, where s-CNTs are exposed to outer interference, the crosstalk can easily penetrate from one CNT to another. This penetration can be reduced by using a dense array of small diameter SWCNTs with (n,m) values (1, 0) and (2, 0) with $d < 2\,\text{nm}$ and having large bandgap. The radial dielectric response of SWCNTs, which is of interest for our analysis, was studied earlier by Benedict et al. (1995). The authors have derived an expression for the dielectric polarizability tensor (α) instead of a dielectric constant. The tensor α_{ZZ} is along the radius of the CNT. Along with that equation, the Thomas–Reiche–Kuhn sum rule is considered which gives the relative dielectric constant ϵ_r of an isotropic medium to derive the relation between α_{ZZ} and ϵ_r obtained as

$$\epsilon_r = 1 + \frac{\alpha_{ZZ} N_e A}{2R\Omega.m} \tag{3.1}$$

where R is the radius of the CNT, E_g is the average bandgap of the CNTs in the bundle, N_e is the number of electrons in a cell that participate for screening and A is the area per atom on the graphite sheet. The radial dielectric constant of a semiconducting SWCNT is calculated as $\epsilon_r = 5$. The coupling capacitance, C_C between two parallel interconnects of radius r, having peripheral s-CNTs of thickness d, and separated by a distance D, can be given as

$$C_C = \frac{2\pi\epsilon}{\ln\left(\frac{D}{2(r-d)}\right) + \sqrt{\frac{D^2}{4(r-d)^2} - 1}} \tag{3.2}$$

after the substitution of ϵ from Eq. 2.28.

Due to large separation between the substrate and interconnect of the order of micrometres (<3 μm), the resulting electrostatic capacitance (C_E) with s-CNTs is at least two orders smaller than the normal C_E of 30 aF (Srivastava et al. 2009). Thus, their contribution to delay is far less and hence is neglected in our SPICE analysis.

3.2.2 *Tentative Process Flow for Integrating S-CNTs*

Size-controlled catalyst nanoparticles were used already by Sato et al. (2006) to fabricate the CNT via structures. In the proposed bundle geometry, the driver of the interconnect must be connected to only the inner metallic CNTs, while the peripheral s-CNTs must be electrically inactive. To this end, the metal contact must be formed such that the central part of the contact contains more metal atoms (thicker) compared to the peripheral part. For 14 nm technology, the nominal dimensions of the interconnect are considered as shown in Fig. 3.1. A mask must be applied to the peripheral part of the contact before metallic CNTs are grown at the centre. Then,

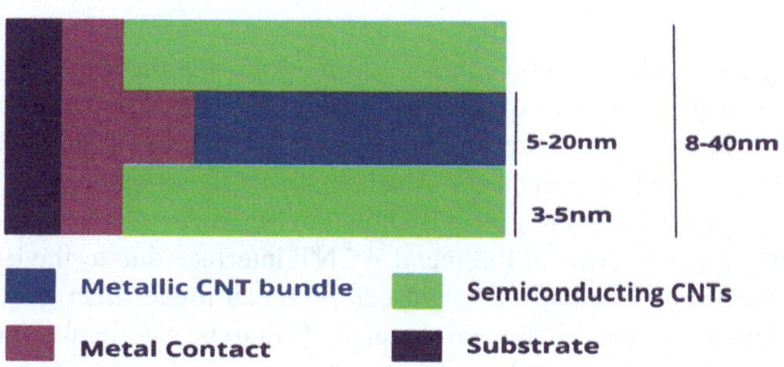

Fig. 3.1 Schematic of proposed geometry of the CNT bundle interconnect with nominal dimensions (Sathyakam et al. 2013)

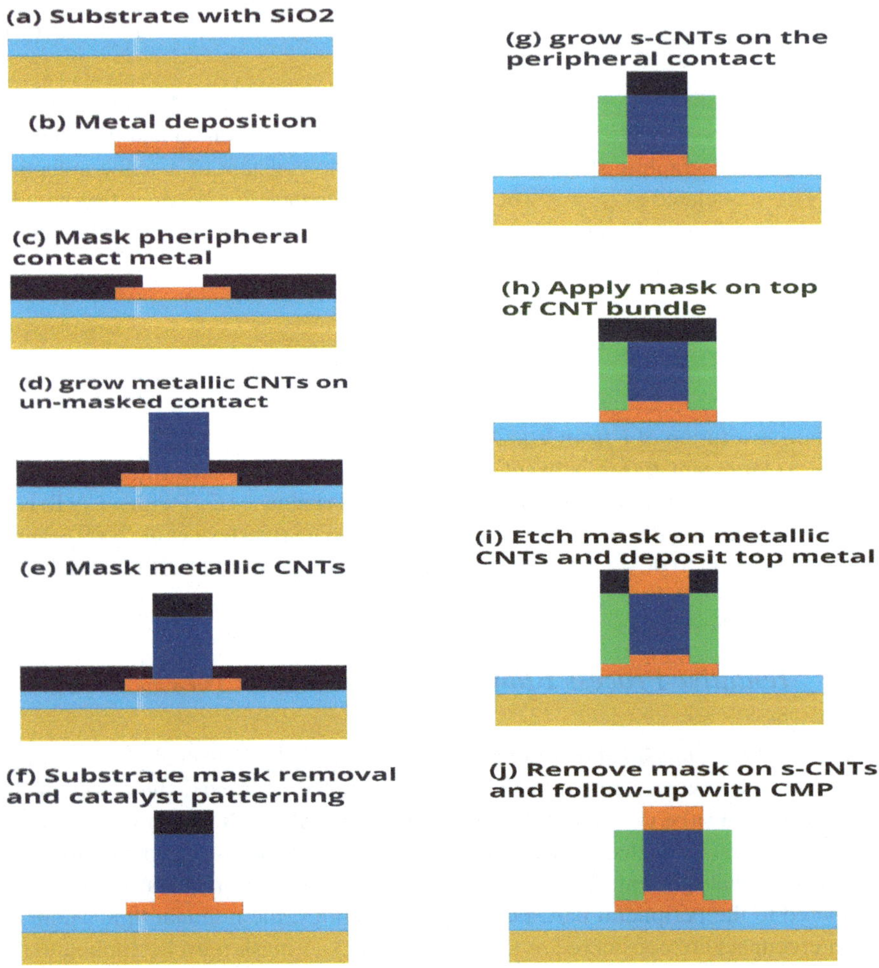

Fig. 3.2 Tentative process flow for proposed CNT interconnect growth (Sathyakam et al. 2013)

by removing the mask from the contact and masking the metallic CNTs, s-CNTs can be grown at the periphery of the interconnect. Chemical mechanical polishing ensures high-density bundles after mask removal. This process is applicable for both vertical and horizontal interconnects. A schematic of the process steps involved in fabricating the proposed interconnect is shown in Fig. 3.2.

A Schottky barrier forms at the metal-s-CNT interface due to the difference in their work functions. The barrier height can be much higher than that of metallic CNT interfaces which causes the peripheral s-CNTs to be electrically inactive. This way, the crosstalk between neighbouring interconnects can be minimized. The central CNT bundle which is metallic in nature provides a perfect path for ballistic conduction. Moreover, the effective wire width of the interconnect is reduced as the s-CNTs fraction in the bundle increases. This in turn reduces the intrinsic quantum as well as the electrostatic capacitance of the interconnect.

3.2.3 MCBs Surrounded by S-CNTs

The single-walled CNTs (SWCNTs) and multi-walled CNTs (MWCNTs) were extensively studied as local, intermediate and global VLSI interconnects. SWCNTs are good for local interconnects, while MWCNTs or double wall CNTs (DWCNTs) are good for global interconnects and via applications (Pu et al. 2009). Recently, CNT bundles that contain both SWCNTs and MWCNTs called mixed CNT bundles are studied. It reveals that MCBs offer less delay in comparison to SWCNT- and MWCNT-based interconnects. Inter-CNT effects in MCBs were studied and hence optimal placement of various CNTs in the bundle was predicted. Earlier works show that MESC is a more effective approach for lesser percentage of error than multiconductor transmission line (MTL) model (Sarto and Tamburrano 2010; D'amore et al. 2010; Giustiniani et al. 2011). This section shows how to reduce the maximum delay due to crosstalk in MCBs by using DWCNTs and semiconducting SWCNTs with the help of MESC approach. MWCNTs play a very important role in MCBs since they can carry more signals at high frequencies as compared to SWCNTs. The shells in an MWCNT are closely bound due to the weak bonds that exist in each CNT shell. So, this makes them more attractive as core conductors than peripheral shields. It is assumed that all shells in an MWCNT are either metallic or semi-metallic in nature. So, at high frequencies, all the shells in an MWCNT conduct current. Thus, it is more appropriate to consider both metallic SWCNTs and DWCNTs at the core and semiconducting SWCNTs at the periphery. Keeping this in mind, a new structure as shown in Fig. 3.3 is proposed where the core has metallic SWCNTs and DWCNTs and the periphery contains semiconducting SWCNTs. Krupke et al. (2004) reported a method to separate semiconducting CNTs from metallic CNTs using their electric field induced polarizabilities. So, the use of semiconducting SWCNTs as peripheral shielding material is emphasized to reduce crosstalk among neighbouring CNT wires. Double-walled CNTs (DWCNTS) were grown and characterized for NEMS applications by Hayashi and Endo (2011). Crosstalk analysis of DWCNT interconnects was done earlier by Pu et al. (2009) and they confirmed that DWCNT bundles are better than SWCNT bundles in terms of performance and delay. Figure 3.4 shows the MESC circuit of the MCB that is considered for analysis. Figure 3.5 shows the schematic of three coupled MCB interconnects. The per unit length (p.u.l) RLC parameters of an MCB were already modelled by the authors of this book. Lee and Lee (2012) reported improved CNT contact treatment after chemical mechanical planarization (CMP) to reduce the contact resistance. They have reported values as low as 85 Ω. Therefore, it is assumed that the contacts are nearly ideal and such small values of resistance will not affect the effective mean free path of the wire. Hence, the inter-CNT capacitance and the inter-CNT tunnelling conductance are not considered in the ESC model.

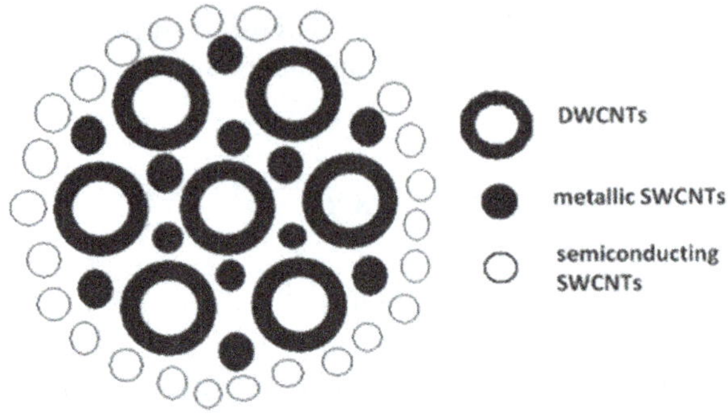

Fig. 3.3 Schematic of mixed CNT bundle proposed for MESC model (Sathyakam et al. 2014)

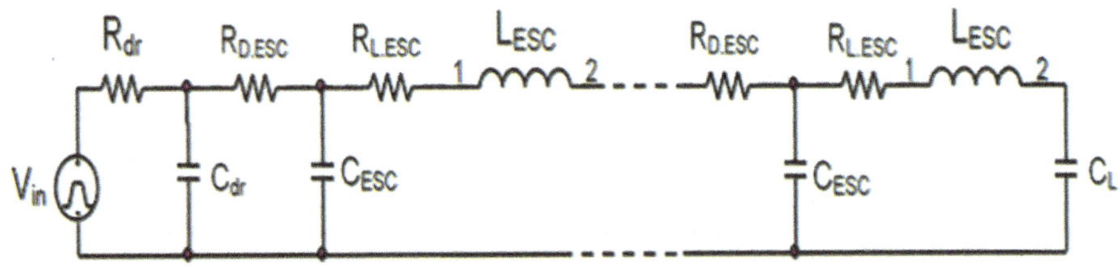

Fig. 3.4 Equivalent single conductor model of MCB (Sathyakam et al. 2014)

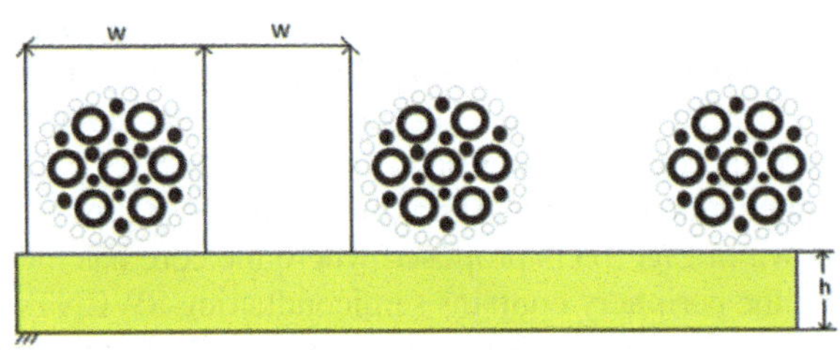

Fig. 3.5 Three coupled MCB interconnects on a substrate (Sathyakam et al. 2014)

3.3 Crosstalk Reduction in MCB Interconnects

3.3.1 Introduction

The operating voltages of high-speed ultra-large-scale integrated (ULSI) circuits are decreasing to subthreshold levels as the integrated chips are down scaled as per the ITRS technology node requirements. The international technology roadmap for semiconductors (ITRS) prepares the future guidelines for manufacturing semiconductor chips. While transistor sizes have decreased to nanometre scale by innovative devices like FinFETs, multi-gate FETs, HEMTs and CNTFETs, it is increasingly becoming difficult to scale interconnects at the same pace. Particularly, scaling of copper interconnects has reached its physical dimensions. Alternative materials such as carbon nanotubes (CNTs) are emerging as nanointerconnects. It is highly desirable that mathematical models of CNTs were formulated to ascertain their behaviour as interconnects. Transmission line modelling of single-walled CNTs (SWCNTs) and multi-walled CNTs (MWCNTs) was developed (Sarto and Tamburrano 2010; D'amore et al. 2010). After initial studies, researchers had proposed that SWCNTs, having diameters from 0.4 nm to a few tens of nanometres, must be used as bundles so that the effective resistance of the CNT bundle will be much lesser compared to the single CNT resistance of 6.45 kΩ/μm. The author of this book had earlier proposed that bundles containing both SWCNTs and MWCNTs in various proportions are ideal as long global interconnects compared to SWCNT bundle and MWCNT interconnects.

Recent work on analysis of CNT bundles ultra-low-power operation at subthreshold conditions reveals that SWCNT bundles are a better choice compared to copper interconnects (Jamal and Naeemi 2011; Pable and Hasan 2012). They had compared the delay, energy per bit and aspect ratio of horizontal SWCNT bundle against mixed CNT bundle and Cu wires. For all cases, CNT wires showed better results than Cu wires. The impact of input bias voltage on the performance of MCBs as intermediate and global interconnects is studied at 20 nm technology node using PTM-HP model-based (www.ptm.asu.edu/) inverter as driver.

3.3.2 Equivalent Circuit Model of MCBs

Two types of mixed CNT bundles are considered here as shown in Fig. 3.6 The bundles have a width of w_B, height h_B and are placed at a distance of h_S from the substrate. The arrangement of CNTs is considered in such a way that in MCB-1, SWCNTs are placed at the core and MWCNTs are placed at the periphery. Alternatively, in MCB-2, MWCNTs are placed at the core and SWCNTs are placed at the periphery. The motivation behind analysing these two structures is to compare the behaviour of these two bundles. Results are shown on the specific behaviour of SWCNTs and MWCNTs on the delay, crosstalk and power dissipation of the inter-

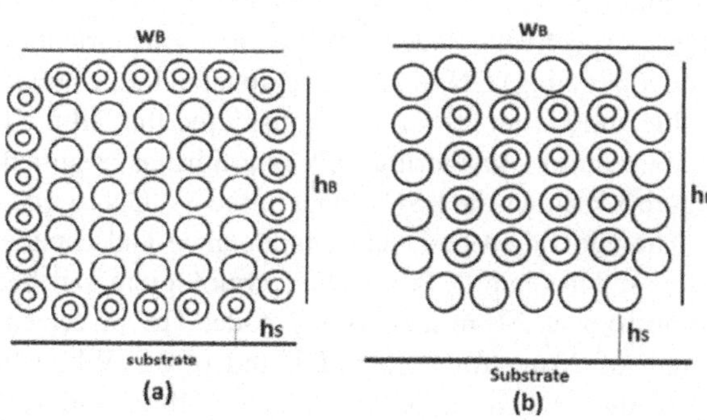

Fig. 3.6 Two possible MCB configurations **a** MCB-1, **b** MCB-2 (Sathyakam et al. 2016)

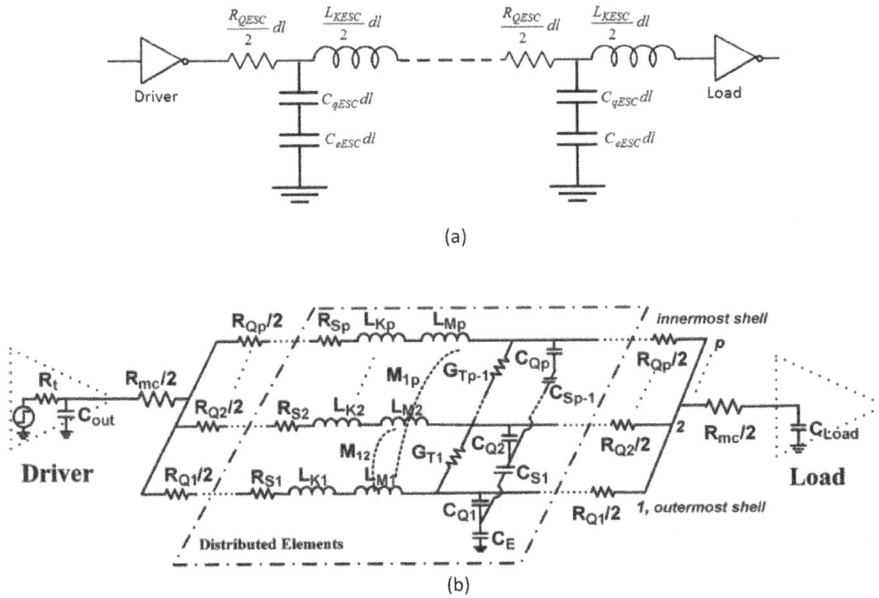

Fig. 3.7 ESC model of **a** SWCNT bundle and **b** MWCNT interconnects

connects in the following section. Earlier, crosstalk and delay analysis was carried out for randomly distributed CNTs in a MCB (Majumder et al. 2012b). Generally, in a mixed CNT bundle, the diameters of the CNTs follow a normal distribution. In both the bundles, it is considered that 25 SWCNTs and 16 MWCNTs are grouped together such that in MCB-1, SWCNTs are arranged as a bundle at the core and is surrounded by the MWCNTs. Alternatively, MCB-2 has 16 MWCNTs arranged as a bundle and is surrounded by 25 SWCNTs, 5 facing each direction. Each SWCNT is of 2 nm diameter and the adjacent CNTs are separated by the van der Waals gap of 0.34 nm. In MWCNTs, only two shells are considered, thus making it a double-walled CNT (DWCNT), for simplicity. The inner CNT has a diameter of 2 nm. The distance between the inner and outer CNT is 0.34 nm and outer shell diameter is 2.68 nm. In MCB-1, 4 MWCNTs and in MCB-2 5 SWCNTs are facing the ground plane, respectively. Recent advancements in fabrication process show that metallic (or semiconducting) CNTs can be segregated into a bundle form using dielectrophoresis (Pable and Hasan 2012). It is considered that all the CNTs in the bundle are metallic, instead of considering only one-third as assumed for all the previous analyses. Figure 3.7 shows the ESC model of the SWCNT bundle and MWCNT interconnects.

The equivalent resistance, inductance and capacitances of the interconnect are discussed in Chap. 2. Predictive technology model-based FETs are used to construct an inverter at 20 nm technology node. The inverter is used as a driver with a gate length of 23 nm for both nFET and pFET. While the nominal voltage is $V_{DD} = 0.9\,\text{V}$, the analysis is performed at subthreshold voltage of 0.3 V in this work. A

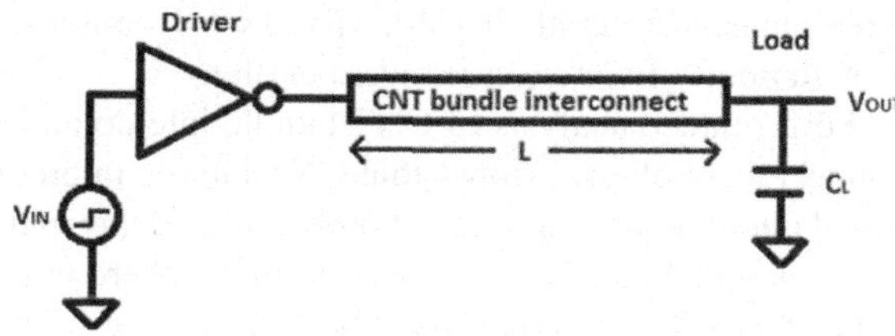

Fig. 3.8 Driver-interconnect-load model of proposed MCB interconnects

Table 3.1 ESC parameters of MCB interconnects

Bundle type	R_{ESC}	L_{kESC}	C_{QESC}	C_{eESC}
MCB-1	56.57 KΩ	0.285 μ H	2.8 pF	32.94 pF
MCB-2	56.57 KΩ	0.285 μ H	2.8 pF	40.25 pF

capacitive load of 10 aF is used. The interconnect length is considered as 500 μm, which is intermediate level wire length as per ITRS. The driver-interconnect-load (DIL) setup that is used for SPICE simulations is shown in Fig. 3.8.

Table 3.1 shows the equivalent single conductor parameters that are calculated for analysis. As can be seen, both the bundles have the same ESC values of R, L and C_Q, as the number of SW and MWCNTs is same. However, for MCB-1, 4 MWCNTs are facing the ground and hence the electrostatic capacitance is 32.94 pF and at the same time, MCB-2 has 5 SWCNTs facing the ground, and hence its electrostatic capacitance is 40.25 pF. The performance analysis and the results are discussed in Sect. 3.5.

3.4 Triangular CNT Bundles as On-Chip Interconnects

3.4.1 Introduction

Scaling down of VLSI circuits to meet Moore's law is continuously posing new challenges to their performance and reliability. New materials and hence new processes are being introduced to meet the requirements of ultra-high-speed and low-power circuit elements in VLSI. Interconnects form a major portion of VLSI chips and hence their performance is of utmost importance. Transistor scaling leads to performance enhancement coupled with low-power operation. However, interconnect scaling leads to performance bottlenecks in terms of delay and power dissipation (Koo et al. 2007). Due to electromigration of copper interconnects at 22 nm technologies and beyond, alternative materials such as carbon nanotubes are being introduced as interconnects. In the last decade, many researchers have been working on modelling, simulation and fabrication of CNT interconnects (Close et al. 2008; Li et al. 2013a). Both metallic SWCNTs and MWCNTs are proposed as local, intermediate and global interconnects (Li et al. 2013b). Mixed CNT bundles, which are more realistic from growth point of view were studied earlier.

Performance analysis of CNT bundle interconnects with copper and other interconnect technologies shows that CNT bundle interconnects outperform other traditional interconnects in terms of propagation delay and power dissipation (Majumder, Kaushik and Manhas 2014). Furthermore, there is no problem of electromigration in CNT bundle interconnects at high frequencies. However, at scaled technologies, it was found that CNT interconnects experience crosstalk that induces a delay in the wires. Since electric signals of very high frequencies (>10 GHz) flow in CNT bundles, an electric field is induced on the adjacent interconnects which causes them to conduct the electric field in the form of a voltage. This phenomenon is called as crosstalk and can be quantitatively modelled using a coupling capacitance, C_C, between the adjacent wires and the induced voltage is called as the crosstalk-induced noise.

Till date, no work has reported alternate geometries for CNT bundle interconnects. Existing works on CNT bundle interconnect modelling use square- or rectangular-shaped bundles as they are easy to fabricate and to meet the W/L aspect ratio requirements stated by the ITRS. However, modification of wire geometry of copper interconnects was performed by Ciofi et al. (2016). They have proposed an interconnect geometry where the line width is decreased by a certain extent to reduce the crosstalk. It can be seen that by adjusting the shape of the CNT bundles, it is possible to keep the inter-wire distance to a minimum, thus reducing the coupling capacitance, and hence the crosstalk-induced delay and power dissipation. This book takes advantage of the fact that CNT bundles can be grown in various geometries depending on the contact pattern geometry. In addition to square or rectangular bundles, it is proposed that CNT bundles of triangular cross section can be used as interconnects. In this section, triangular CNT (T-CNT) bundle interconnects are proposed for very high-speed VLSI circuits.

3.4.2 Triangular CNT Bundle Geometry-1

Square-type CNT bundles (height/width aspect ratio = 1) were found to be better structures than rectangular ones (AR = 1.5, 2) as the former has a smaller number of CNTs in adjacent bundles that are coupled to each other (Majumder, Kaushik and Manhas 2014). Square bundles are considered as reference structure for comparison with T-CNT bundles. As depicted in Fig. 3.9, the bundle is placed at a distance of h_t from the ground plane. The number of CNTs along the width (W) is n_W and the number of CNTs along the height (H) of the bundle is n_H. Each CNT has a diameter d of 1 nm. Triangularly arranged CNTs can be formed into bundles by bottom-up fabrication approaches which will be discussed elsewhere. It can be seen that an equilateral triangular arrangement gives the optimum distance between CNTs in adjacent bundles thereby reducing their coupling.

The T-CNT bundle has a total number of CNTs in the bundle given as

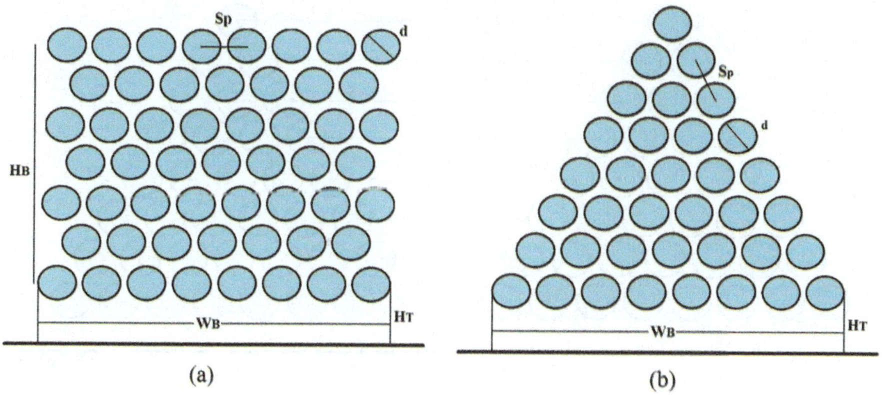

Fig. 3.9 Geometry of **a** square and **b** triangular CNT bundle interconnects (Sathyakam and Mallick 2019)

$$n_{TCNT} = \sum_{b=1}^{n} n_b(b) = \frac{n_b(n_b + 1)}{2} \quad (3.3)$$

where n_b is the number of CNTs at the base of the bundle. It is assumed that the width, W_B of all the CNT bundles is the same so that the electrostatic capacitance between the bundle and substrate remains same. Here, a square CNT bundle with $W_B = H_B = 34$ nm with 564 CNTs in it has been considered. T-CNT bundle is having the same width with 300 CNTs in it. Now, the coupling capacitance of T-CNT bundle interconnects is modelled when two wires are placed adjacently on a substrate as shown in Fig. 3.10. For two capacitively coupled T-CNT bundles that have n CNTs along their side, the coupling capacitance of the nth corresponding CNT pair is

$$C_C^{CNT} = \frac{2\pi\epsilon_r}{cosh^{-1}(y_n/d)} \quad (3.4)$$

where $y_n = y_{n-1}S_p$ is the inter-CNT distance of the nth pair of coupled CNTs, $n = 1, 2, 3, 4, \ldots$ and $y_0 = 34$ nm. The total coupling capacitance C_C of the bundles can be given as

$$C_C = \sum_{m=1}^{n} C_C^{CNT}(m) \quad (3.5)$$

The C_C values for T-CNT bundles are computed using (3.5) for crosstalk analysis of coupled interconnects.

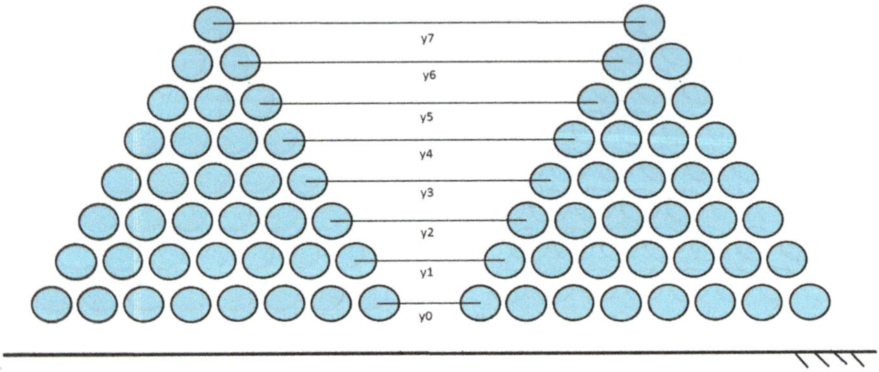

Fig. 3.10 Coupled T-CNT bundle interconnects (Sathyakam and Mallick 2019)

Table 3.2 ITRS 2015 data for 20 nm technology node interconnects

Parameter	Intermediate wires	Global wires
Min. Wire pitch (nm)	34	51
Min. half pitch (nm)	16	26
Thickness, H (nm)	100	125
Aspect ratio (W/L)	2	2.34
V_{DD} (volts)	0.7	0.8

3.4.3 Triangular CNT Bundle Geometry-2

In coupled triangular CNT bundle interconnects, the inter-CNT distance varies optimally while maintaining the overall bundle distance within the ITRS requirements as enlisted in Table 3.2.

For square-type CNT bundles as shown in Fig. 3.11a, the number of CNTs along the width is n_W and the number of CNTs along the height is n_H. The total number of CNTs, n_{SB}, is given by

$$n_{SB} = n_W \times n_H \tag{3.6}$$

Hence, the width w_B and height h_B of the bundle are given as

$$w_B = (n_W \times D) + (n_W - 1) \times \delta \text{ and } w_H = (n_H \times D) + (n_H - 1) \times \delta \tag{3.7}$$

where δ is the van der Waals gap equal to 0.34 nm. The arrangement of CNTs is considered in columns rather than the one described in the previous section such that the total number of CNTs in square-type as well as triangle-type bundle is same, making the analysis easy and comparable. Moreover, the CNT bundle model can be practically realized by a bottom-up approach, where CNTs can be deposited on a substrate by dielectrophoresis method and metal contacts can be fabricated on either

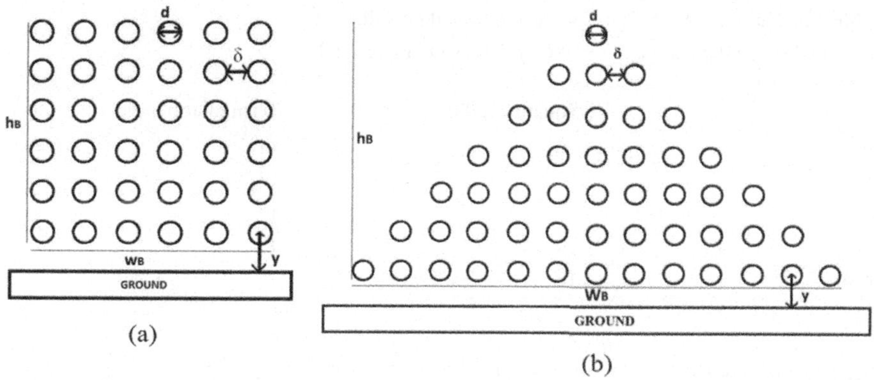

Fig. 3.11 CNT bundle geometries considered in this section. **a** Square CNT bundle and **b** Triangular CNT bundle (Sathyakam et al. 2020)

sides. For the triangular bundle as shown in Fig. 3.11b, considering the fact that the number of CNTs along each row is odd, the total number of CNTs in the bundle n_{TB} is

$$n_{TB} = (n_S \times n_B) - \lfloor n_S \times (n_S - 1) \rfloor \tag{3.8}$$

where n_S is the number of CNTs along the side and n_B is the number of CNTs along the base of the triangular bundle. For the sake of simplicity, it is considered that $n_{SB} = n_{TB}$. Table 3.3 gives the number of CNTs in the bundles and the number of CNTs facing the ground. The resistance, capacitance and inductance of the CNT bundle interconnects are discussed in Chap. 2. The resulting RLC values for square and triangular bundles of Fig. 3.11 are listed in Table 3.4. The two coupled lines are characterized as aggressor and victim lines. The coupling capacitance, C_C, is given as

$$C_C = \frac{2\pi\epsilon}{cosh^{-1}(y_n/d)} \tag{3.9}$$

where d is the diameter of the CNT and y_n is the centre-to-centre distance between corresponding CNTs in aggressor and victim lines. The distance between corresponding CNTs in coupled triangular bundles is given as

$$Y_n = 2(d + \delta) + Y_{n-1} \tag{3.10}$$

where n = 1, 2, 3, …, d is the diameter of the CNT and δ is the van der Waals gap of 0.34 nm. Fabrication of the proposed triangular CNT bundles can be achieved by forming V-shaped groves on a SiO_2 substrate by lithography process. Metal electrodes which can be made of gold/tungsten must be formed on the either sides of the groves to guide the dielectrophoresis assembly of CNTs between them (Close and Wong 2007). The substrate must be placed in a solution such as dichlorobenzene

Table 3.3 Number of CNTs in square and triangle bundles

No. of CNTs in the bundle ($n_{SB} = n_{TB}$)	No. of CNTs facing ground in	
	Square bundle	Triangular bundle
36	6	11
49	7	13
64	8	15
81	9	17
100	10	19

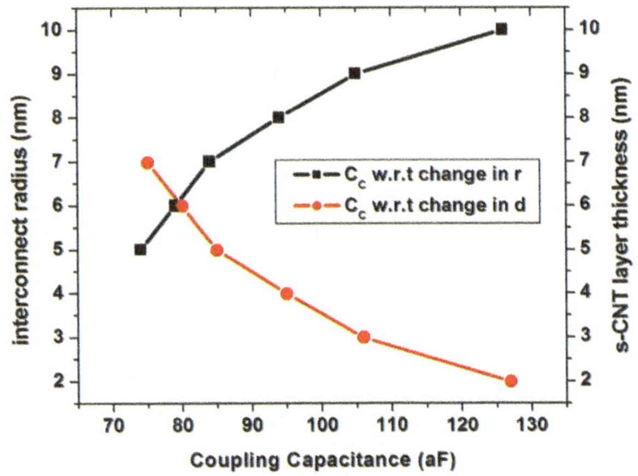

Fig. 3.12 Variation of capacitance with increase in thickness of the s-CNT layer and wire radius for different s-CNT layer thicknesses (Sathyakam et al. 2013)

Table 3.4 ESC parameters

Length (μm)	R_{ESC} (kΩ)	L_{ESC} (μH)	C_{QESC} (pF)	C_{eESC} (fF)	
				Square	Triangle
500	89.580	0.111	7.2	90	165
1000	179.16	0.222	14.4	180	330
1500	268.74	0.333	21.6	270	495
2000	358.32	0.444	28.8	360	660

where CNTs can be dispersed in it. An alternating voltage of 20 V peak-to-peak must be applied at 500 kHz between the electrodes. CNTs align themselves neatly between the electrodes. Finally, top metal contact can be deposited at either ends of the CNT bundles and remaining contact metal, if any, can be etched away. As the trenches are in 'V' shape, the CNT bundle geometry will be 'V' shaped or an inverted triangular.

Table 3.5 Reduction of coupling capacitance and crosstalk-induced delay for various s-CNT layer thickness values

s-CNT layer thickness, d (nm)	C_C with s-CNTs (aF)	Reduction of C_C with s-CNTs (%)	Delay with s-CNTs (ns)	Delay without s-CNTs (ns)	Reduction of delay with s-CNTs (%)
2	144.5	50.6	2.5	2.72	8.08
3	113.0	61.4	2.495	2.60	4.03
4	89.0	69.6	2.494	2.643	5.63
5	69.5	76.2	2.493	2.675	6.8
6	51	82.5	2.492	2.721	8.41

3.5 Results and Discussions

This section details the results of crosstalk and noise analysis of various types of CNT bundle interconnects discussed in the previous sections.

3.5.1 Analysis of S-CNT-Surrounded CNT Interconnects

Figure 3.12 shows the decrease of capacitances calculated for increase in s-CNT layer thickness and interconnect radius. From Eq. 3.5, the values of C_C are computed with and without peripheral s-CNTs and compared in Table 3.5. The coupling capacitance is reduced from 50.6 to 82.5%. The capacitance without an s-CNT layer is 292.89 aF for interconnect separation of 15 nm and interconnect radius of 7 nm. Reduction in C_C after incorporating s-CNTs in the bundle is shown in Table 3.5. The reduction in capacitance ranged from 93.8 to 97.7% when both wire radii (from 9 to 19 nm) and distance between wires (from 10 to 20 nm) are increased, indicating the impact of s-CNTs over a range of dimensions. A HSPICE-based analysis is carried out to find the crosstalk-induced propagation delay of the proposed interconnect system. A CNT bundle with suitable driver and load values is considered. An equivalent single conductor model depicted in Fig. 3.13a and an aggressor–victim–aggressor–victim interconnect system shown in Fig. 3.13b are considered for the delay analysis. HSPICE analysis shows that delay was reduced when s-CNTs were used at the periphery. The delay values and the corresponding delay reduction are shown in Table 3.5. EM simulations were carried out to ascertain the EM wave blocking nature of s-CNTs. Moreover, s-CNTs with large bandgap and higher barrier height at the contact interface ensure that electrical conduction does not occur in s-CNTs. The crosstalk was measured for both in-phase and out-of-phase switching scenarios. The wire length is varied from 1 to 1000 μm and a voltage of 0.1 V is applied to all lines. In both cases, the crosstalk between victim and aggressor lines is evident in Fig. 3.14a where there are no peripheral CNTs. However, Fig. 3.14b shows the current

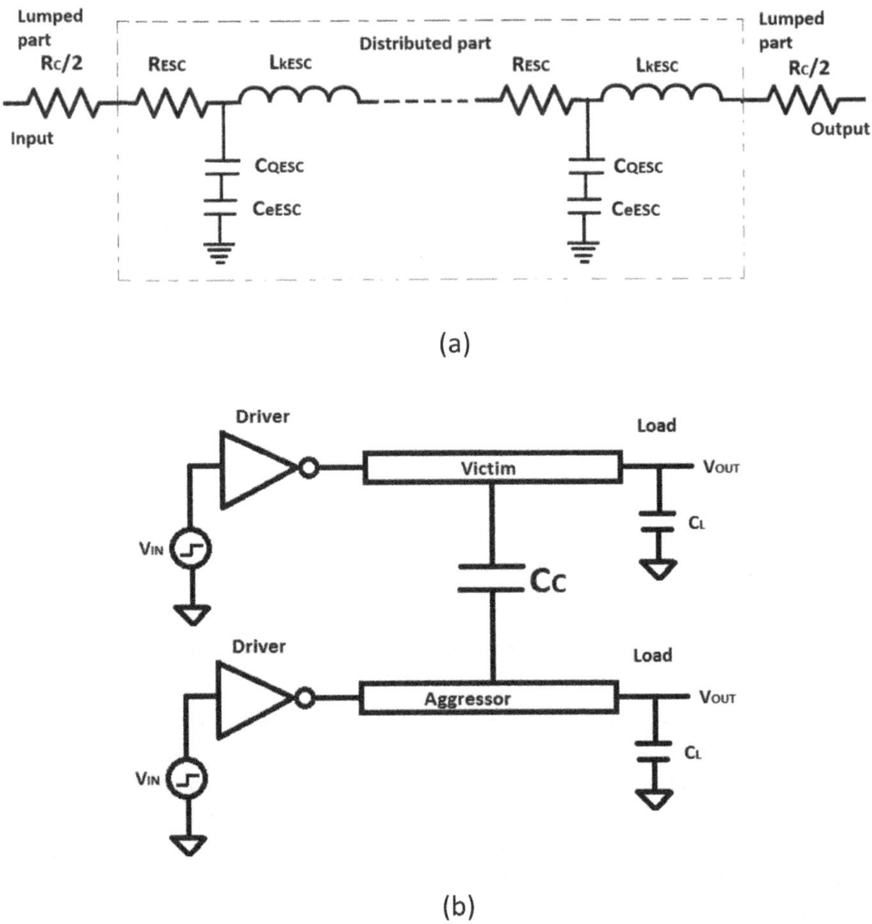

Fig. 3.13 **a** Equivalent circuit model considered for HSPICE analysis. **b** Aggressor–victim–aggressor model including C_C

distribution with peripheral s-CNTs that reveals the absence of interference between the two interconnects where the current distribution was found to be maximum at the metallic CNT part of the interconnect and it fades away as the outer s-CNT thickness increases. To summarize, this section highlighted the role of s-CNTs as EM wave blockers in CNT interconnect systems. The modified capacitance expression after including the s-CNT thickness is shown and the coupling capacitance values are derived. The Comsol EM multi-physics and HSPICE simulations were shown to ascertain the EMI-blocking property of s-CNTs. Interconnect delay values were calculated and compared with and without s-CNTs. It was found that delays reduced when s-CNTs were used at the periphery of the CNT interconnect. The fabrication of the proposed structure and characterization are future works.

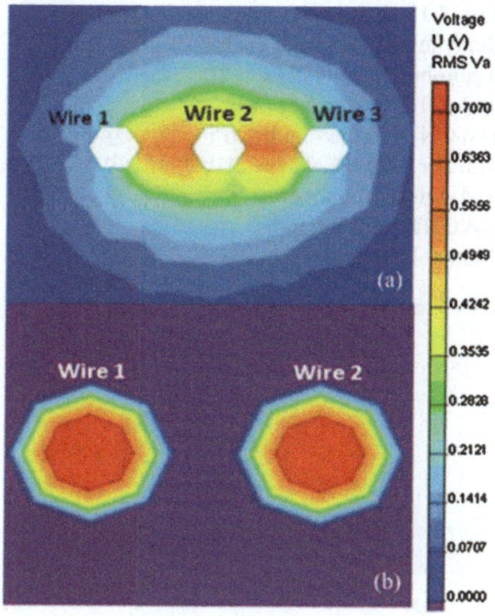

Fig. 3.14 Current distribution for **a** tri-interconnect system and **b** two interconnect system with peripheral s-CNTs (Sathyakam et al. 2013)

3.5.2 Analysis of S-CNT-Surrounded MCB Interconnects

One of the main advantages of using semiconducting CNTs at the periphery is to reduce the coupling capacitance effects between adjacent lines. Also, the electrostatic capacitance between the wire and the substrate can be reduced by placing the right amount of semiconducting CNTs at the periphery. After considering appropriate driver, load resistance and capacitance values, a pulsed voltage is applied to the interconnect circuit. Figure 3.15 shows the comparison of length versus crosstalk-induced time delay for various counts of SWCNTs and DWCNTs in the MCB. The parameters according to the ITRS-2013 interconnect report for 22 and 14 nm technology node were considered.

Crosstalk-induced delay for interconnects of 100 to 1000 μm is analysed for a fixed number of SWCNTs and varying the count of DWCNTs in the bundle. The same is done by varying the count of SWCNTs with constant number of DWCNTs. It is observed that as length increases, the delay increases. The delay has been increased drastically in those bundles that have a smaller number of CNTs. The crosstalk-induced delay for the best and the worst cases is shown along with the delay values of other MCBs.

Table 3.6 shows the delay values for the best and worst cases along with the improvement in delay among the two cases. It is observed that MCBs with more DWCNTs show least delay compared to that of more SWCNTs at the core. The reason is that semiconducting CNTs are structurally dissimilar with respect to metallic CNTs and oppose electron tunnelling into the inner tubes as the weak bond interaction is absent here. Thus, the overall delay of the wire is reduced. It is evident that the best-

Fig. 3.15 Crosstalk-induced delay at different lengths for **a** number of SWCNTs = 10 and different number of DWCNTs in the bundle, **b** number of DWCNTs = 10 and different number of SWCNTs in the bundle (Sathyakam et al. 2014)

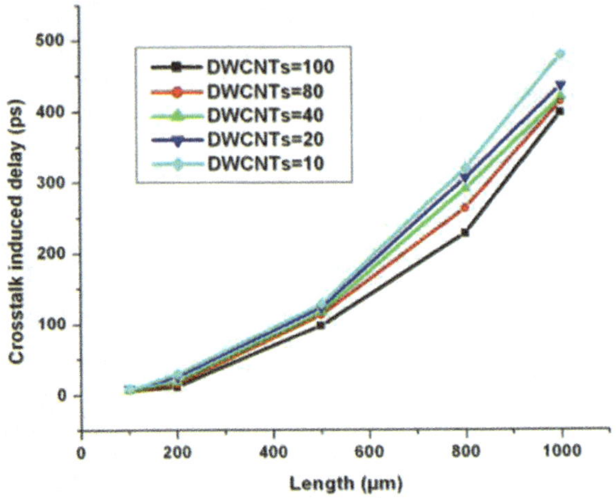

(a)

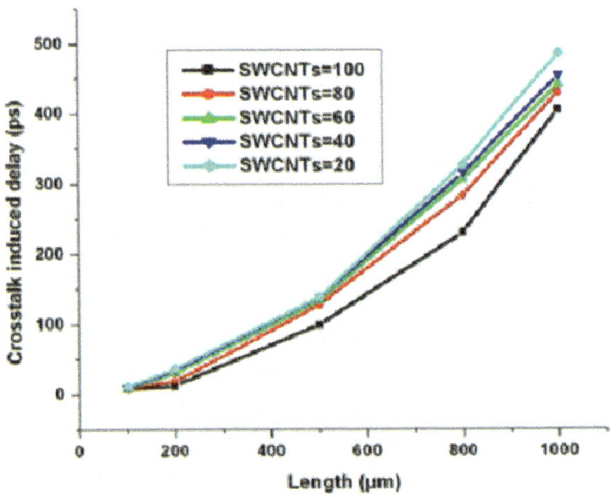

(b)

case MCBs show the best performance improvement at 200 μm. It means they are suitable as short intermediate interconnects at 14 nm technologies.

The improvement in delay for the best as well as the worst cases of the proposed models, when compared to other MCB structures from Majumder et al. (2012b), is presented in Table 3.7 . It is observed that, the best case presented in this book fare better indicating that they are ideal as local and intermediate interconnects when compared to the structure by Majumder et al. (2012b).

Table 3.6 Improvement in delay for the best-case and worst-case delays in the MCB for various CNT counts

Length (μm)	Crosstalk induced delay (ps)		Improvement in delay (%)
	SWCNTs = 10, DWCNTs = 100 (best case)	SWCNTs = 10, DWCNTs = 10 (worst case)	
100	7	11	36.37
200	11	37	70.28
500	97	139	30.22
800	227	328	30.8
1000	398	486	18.11

Table 3.7 Improvement of crosstalk-induced delay for various configurations of MCBs

Length (μm)	Crosstalk-induced delay (ps); * - Majumder et al. (2012b)		Improvement in delay for our best case w.r.t best case in *	Improvement in delay for our worst case w.r.t best case in *	Improvement in delay for our worst case w.r.t worst case in *
	Best-case MCB delay in *	Worst-case MCB delay in *			
100	14	29	50	21.43	62.07
200	31	67	64.52	16.22	44.78
500	138	321	29.72	0.72	56.7
800	331	797	31.42	0.91	58.85
1000	504	1225	21.04	3.58	60.33

The delay is smaller in our worst-case MCBs as compared to the best case in Majumder et al. (2012b) which it is due to the presence of DWCNTs at the core of the proposed structure in this book.

Comparison of the proposed MCB structures with MWCNTs with various shell counts is shown in Table 3.8. It is evident that delay values of MCBs are lesser in one to three orders of magnitude than that of MWCNT interconnects in Majumder et al. (2012a).

This can be attributed to the enhanced EMI protection that semiconducting SWCNTs offer to our structure. Also, the improved contact treatment and hence the reduced contact resistance of DWCNTs contribute to the reduction of delay as opposed to MWCNTs which have multiple shell contacts.

Table 3.8 Comparison of delay values between MWCNTs and MCBs

Length (μm)	Crosstalk-induced delay (ps); ∗ - Majumder et al. (2012a)				
	SWCNTs = 10, DWCNTs = 100 (best case)	SWCNTs = 10, DWCNTs = 10 (worst case)	MWCNTs (no. of shells = 10) in ∗	MWCNTs (no. of shells = 20) in ∗	MWCNTs (no. of shells = 30) in ∗
100	0.007	0.011	0.302	0.265	0.252
200	0.011	0.037	0.390	0.311	0.283
500	0.097	0.139	0.782	0.468	0.377
800	0.227	0.328	1.476	0.656	0.493
1000	0.398	0.486	2.176	0.827	0.572

Table 3.9 Performance analysis of proposed interconnects

	Propagation Delay (μ s)	Power dissipated (μ W)	Power-delay product (PDP)
MCB-1	1.71	1.6670	2.85057
MCB-2	1.82	1.6672	3.034304

3.5.3 Performance Comparison of MCB Interconnects

Transient analysis is performed by applying a pulsed voltage of 0.3 V and a frequency of 1 GHz. The propagation delay, power dissipated and hence the power-delay product (PDP) of MCB-1 and MCB-2 are shown in Fig. 3.6. Table 3.9 summarizes the obtained values.

It can be seen that the propagation delay is 0.09 μs lesser for MCB-1 as it has lesser electrostatic capacitance than MCB-2. This is due to the less number of CNTs at the periphery for MCB-1. Secondly, the power dissipated by MCB-1 is lesser than MCB-2 by 0.2 nW for the same reason. So, obviously, depending on the PDP, MCB-1 shows a better performance than MCB-2. Next, crosstalk analysis of capacitively coupled MCBs is performed as shown in Fig. 3.16.

First, the coupling capacitance between two adjacent MCBs of the same type is computed for a length of 500 μm. Transient analysis is performed by applying a pulsed voltage of 0.3V for both coupled wires. Due to the coupling capacitance, the crosstalk-induced delay will raise. This delay at in-phase and out-of-phase scenarios is found out. Figure 3.17 shows the out-of-phase output for MCB.

SmartSPICE analysis reveals that the crosstalk-induced delay, power dissipated and hence the PDP of MCB-2 are slightly higher than MCB-1 which makes MCB-1 more attractive as an interconnect at intermediate level as shown in Table 3.10. In conclusion, two different mixed CNT bundle structures with different arrangements of CNTs are modelled and their performance was simulated. MCB-1 and MCB-2 were modelled using a driver-interconnect-load setup. Transient analysis using

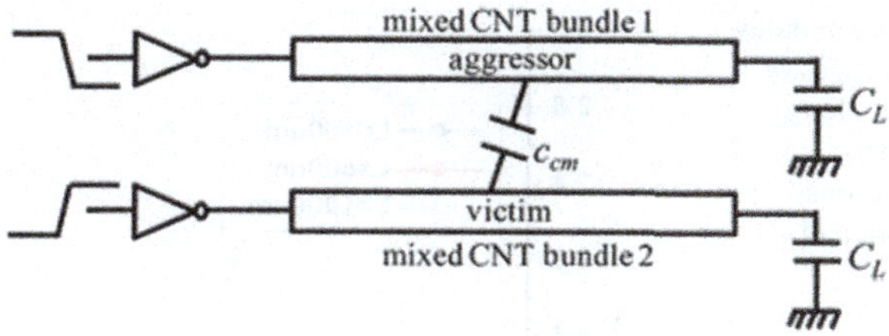

Fig. 3.16 Schematic of two capacitively coupled interconnects

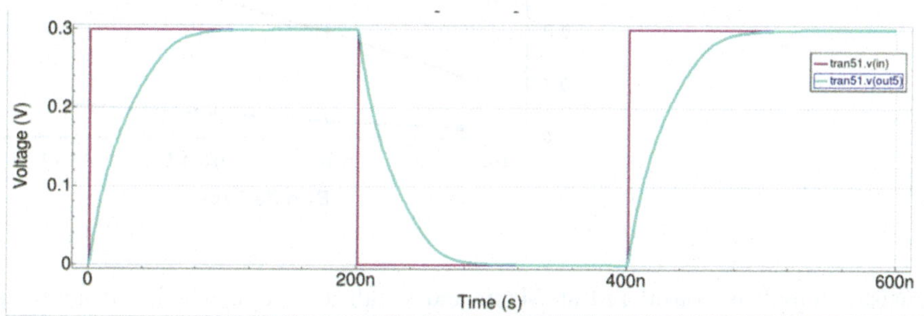

Fig. 3.17 Crosstalk analysis of out-of-phase scenario **a** input pulse, **b** output pulse (Sathyakam et al. 2016)

Table 3.10 Crosstalk analysis of proposed interconnects

	Crosstalk delay (μs)	Power dissipated (μW)	Power-crosstalk delay product (PCDP)
MCB-1	0.35	3.0394	1.06379
MCB-2	0.38	3.0396	1.155048

SmartSPICE shows that MCB-1 has lesser propagation delay than MCB-2 by 0.11 μs and power dissipated by MCB-1 is lesser than MCB-2 by 0.02 nW. Further, crosstalk analysis of both bundles shows that MCB-1 is having lesser crosstalk-induced delay than MCB-2. The ultimate performance is indicated by the PDP, which is lower for MCB-1. So, MCB-1 is preferred over MCB-2 as intermediate level interconnect at subthreshold voltages.

3.5.4 Analysis of Triangular CNT Bundle Interconnects

To analyse the behaviour of the proposed CNT bundles, a pulsed voltage of 0.9 V, 4 μs pulse width, 8 μs pulse duration, 0.1 μs rise time and fall time is applied to the ESC-modelled SPICE circuits. An inverter made of PTM-HP FinFET 20 nm technology models is considered for simulations using Silvaco SmartSPICE. The

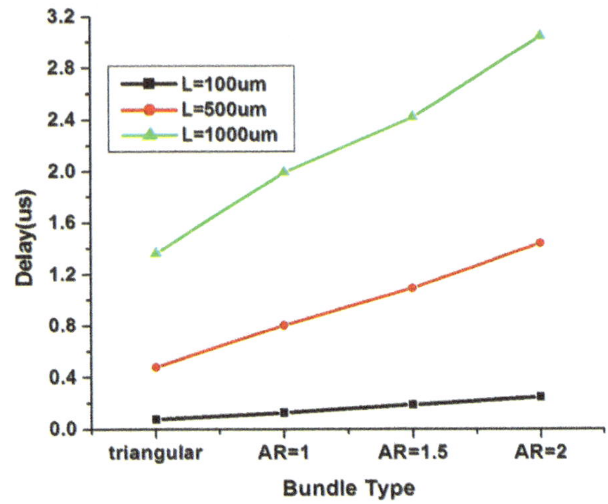

Fig. 3.18 Propagation delay for various bundle types of AR = 1, 1.5 and 2, along with T-CNT bundle interconnects at various lengths (Sathyakam and Mallick 2019)

transistor length is considered at 24 nm and width at 15 nm, which are nominal values. Figure 3.18 shows the propagation delay of the proposed T-CNT bundles along with bundle of AR = 1, 1.5 and 2.

It can be seen that the propagation delay is least for T-CNT bundles compared to bundles of AR = 1, 1.5 and 2. From this, it can be inferred that the optimum number of CNTs in a bundle is the same as the number of CNTs in T-CNT bundles that gives optimum delay values for a given length. This is because each CNT contributes to the intrinsic quantum resistance of 6.45 $k\Omega/\mu m$ and quantum capacitance of 400 $aF/\mu m$. So, in order to achieve minimal delay, both R and C should be as low as possible. So, typically, in this case, the T-CNT bundle quantum capacitance is smaller by 46% compared to AR = 1 bundle. Next, the power dissipation by each type of bundles is explained. From Fig. 3.19, it can be observed that the power dissipated by T-CNT bundles is far lesser than its counterparts. When compared to its closest competitor, the power dissipated by T-CNT bundles is 56% lesser than AR = 1 bundles at 1000 μm. So, this demonstrates the validity and effectiveness of using T-CNT bundles as local, intermediate and global interconnects.

Crosstalk analysis of the proposed interconnects is performed where one wire is excited by a pulse signal and another wire is held at logic high. Figure 3.20 shows the model of two interconnects placed on a substrate at a distance of H_T that is used for the analysis. The crosstalk-induced delay in the wires is maximized when the wires are excited with opposite polarity pulses. However, this book shows the performance comparison of T-CNT bundles versus square bundles. The impact of increasing the inter-wire distance on the performance of CNT interconnects is analysed. So, the crosstalk-induced delay and power dissipation for inter-wire distances of S = W, S = 1.5 W and S = 2 W are found out. As can be seen from Table 3.11, the crosstalk-induced delay increases for increase in length. Further, as the spacing S is increased, the delay is reduced. When compared to square bundles, T-CNT bundles have 40%,

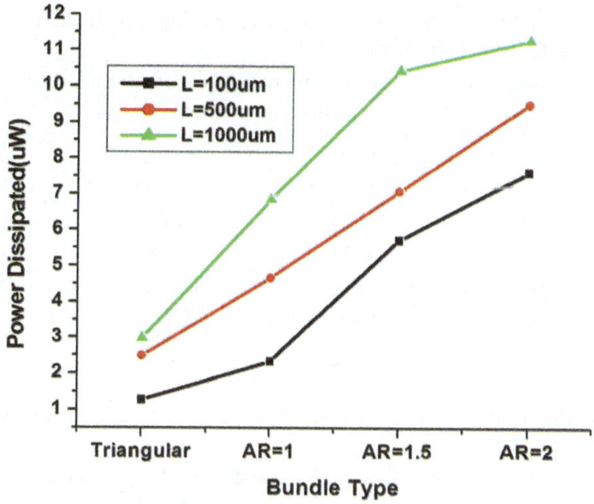

Fig. 3.19 Power dissipated for various bundle types of AR = 1, 1.5 and 2 along with T-CNT bundle interconnects at various lengths (Sathyakam and Mallick 2019)

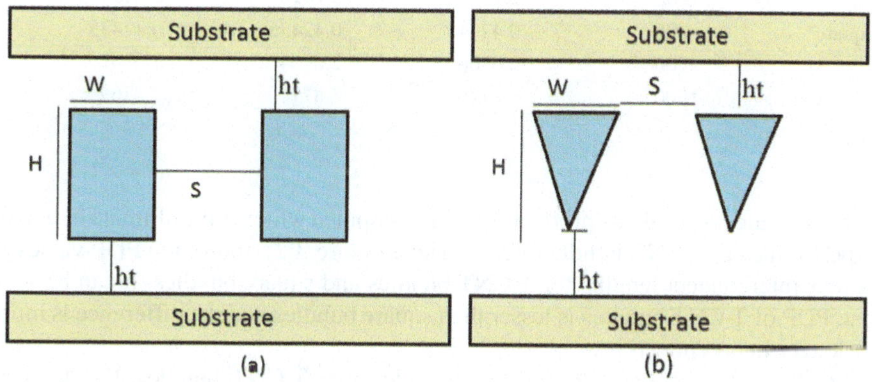

Fig. 3.20 Two coupled **a** square and **b** triangular CNT bundle interconnects placed on a substrate (Sathyakam and Mallick 2019)

41% and 33% reduction in delay for 100, 500 and 1000 μm lengths when S = W, respectively.

From Table 3.12, it can be seen that the trend in power dissipation is that as the length increases, power dissipation increases, whereas if the wire spacing is increased, power dissipation decreases. Moreover, T-CNT bundles dissipate less power than square bundles. It is 41%, 45% and 44% reduction in power dissipation by T-CNT bundles compared to square bundles at 100, 500 and 1000 μm lengths for S = W, respectively.

Table 3.11 Crosstalk-induced delay for various wire spacings for triangular and square CNT bundles, *(Majumder, Kaushik and Manhas 2014)

Geometry	Length (μm)	Crosstalk delay for different interconnect spacings (μs)		
		(S = W)	(S = 1.5 W)	(S = 2 W)
Triangular	100	0.10124	0.10123	0.10122
	500	0.51638	0.5163	0.51627
	1000	1.3881	1.3879	1.3878
Square*	100	0.17147	0.17146	0.17144
	500	0.89019	0.89010	0.89004
	1000	2.0941	2.0938	2.0937

Table 3.12 Power dissipation for various wire spacings for triangular and square CNT bundles, *(Majumder, Kaushik and Manhas 2014)

Geometry	Length (μm)	Power dissipated for different interconnect spacings (μs)		
		(S = W)	(S = 1.5 W)	(S = 2 W)
Triangular	100	0.22515	0.22512	0.2251
	500	1.1308	1.1309	1.1310
	1000	2.1629	2.1626	2.1624
Square*	100	0.42441	0.42436	0.42433
	500	2.0706	2.0704	2.0702
	1000	3.8776	3.8772	3.8769

Finally, the power-delay product (PDP) is computed which is the ultimate indicator of performance of CNT bundle interconnects. Figure 3.21 shows the PDP values at various interconnect lengths for T-CNT bundles and square bundles. It can be seen that, PDP of T-CNT bundles is lesser than square bundles and the difference is more for longer interconnects.

So, from this analysis, it can be concluded that T-CNT bundles, due to their reduced coupling capacitance, perform much better than traditionally used square CNT bundles.

Now, another approach of modelling triangular CNT bundles is analysed. A driver-interconnect-load (DIL) setup is used to perform transient analysis. After considering the appropriate ESC parameters for the interconnects, transient analysis is performed by applying a pulsed signal to the DIL setup. From the results shown in Fig. 3.21, it can be seen that the electrostatic coupling capacitance, even though an order of magnitude lesser than the quantum capacitance, impacts the propagation delay of the interconnects. The simulations are carried out considering 36 CNTs, each CNT of 2 nm diameter and at room temperature. In square bundle, the number of CNTs facing ground is 6, while for triangular bundle, it is 11.

The propagation delay is slightly higher for triangular interconnects compared to square interconnects (Fig. 3.22). Transient waveforms of the output voltages at 500,

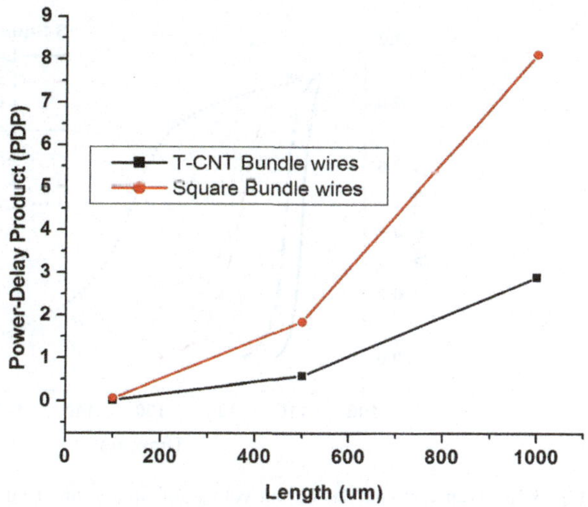

Fig. 3.21 Power-delay product of T-CNT and square bundle wires at various lengths (Sathyakam and Mallick 2019)

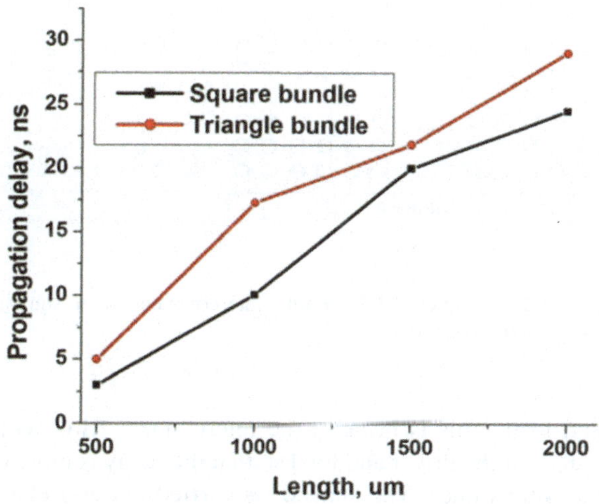

Fig. 3.22 Propagation delay of square-type and triangle-type bundled CNT interconnects (Sathyakam et al. 2020)

1000, 1500 and 2000 μm lengths are shown in Fig. 3.23. The difference in delay between square and triangular bundles is 4 ns at 500 μm, 10 ns at 1000 μm, 13 ns at 1500 μm and a negative difference of 15 ns at 2000 μm. The main reason behind the smaller peak voltage for triangular bundle at 2000 μm is the contribution of length-dependent quantum capacitance of 28.8 pF and the electrostatic capacitance of 660fF, which is highest among other lengths.

Two coupled interconnects are used to analyse the crosstalk effects as shown in Fig. 3.24. The wires are energized using a 1 V pulsed signal for analysis. Various switching scenarios, viz. in-phase and out-of-phase are considered. The load capacitance of $C_L = 10\,\text{aF}$ is considered which corresponds to the minimum-sized driver

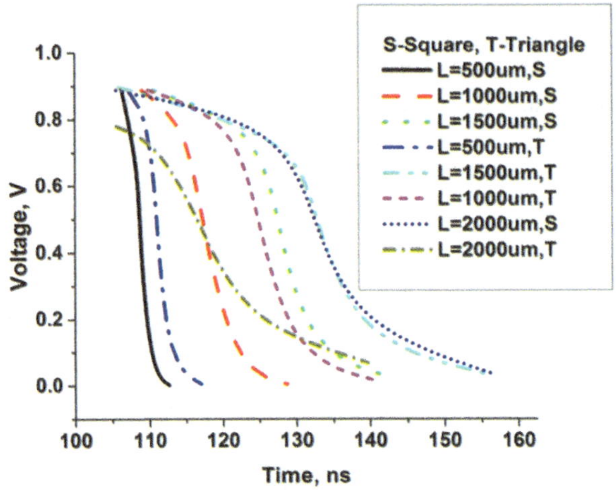

Fig. 3.23 Transient analysis output voltage of square- and triangle-type bundles at different interconnect lengths of 500, 1000, 1500 and 2000 μm (Sathyakam et al. 2020)

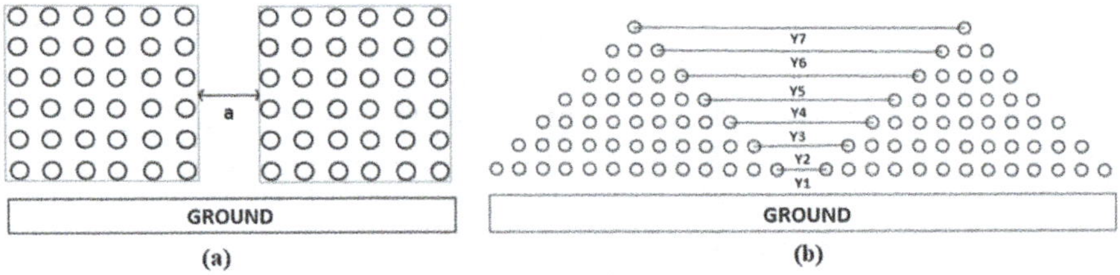

Fig. 3.24 Coupled CNT bundle interconnects of **a** square geometry, **b** triangle geometry (Sathyakam et al. 2020)

and load. This is because the interconnect RLC will be dominating over the RC values of the driver and load so that the delay is more dependent on the interconnect parasitic values. The analysis is carried out at global interconnect lengths of 500, 1000, 1500 and 2000 μm. The results of delay analysis are shown in Fig. 3.25.

Figure 3.26 shows the crosstalk-induced delay of both square- and triangle-type CNT bundle interconnects at lengths of 500, 1000, 1500 and 2000 μm. It is found that the crosstalk-induced delay is less for triangle-type bundle interconnects at all lengths. The least possible delay improvement of triangle-type bundles over square-type bundles is 19.6% for 36 CNTs in a bundle and the maximum improvement was 29.5% for 100 CNTs in a bundle as shown in Fig. 3.27.

The main reason for this improvement can be attributed to the reduction in the coupling capacitance between the adjacent interconnects in triangular CNT bundle interconnects. Also, the increase in delay improvement for larger bundles is due to the fact that more number of CNTs are farther apart and less coupled than for the

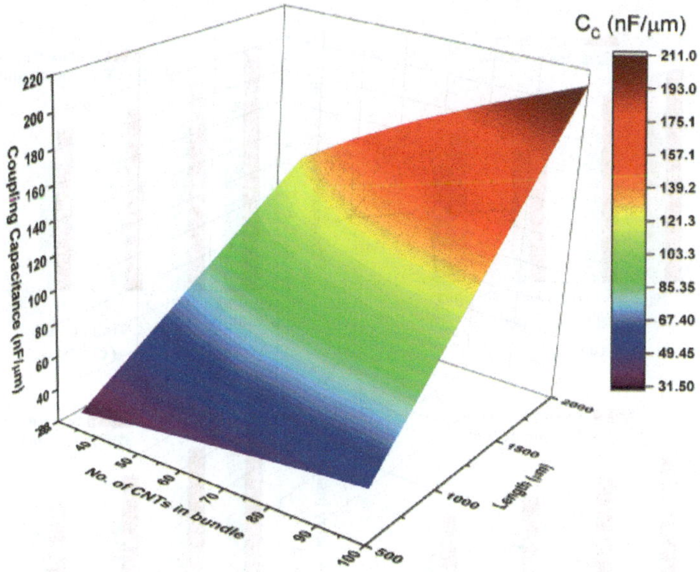

(a)

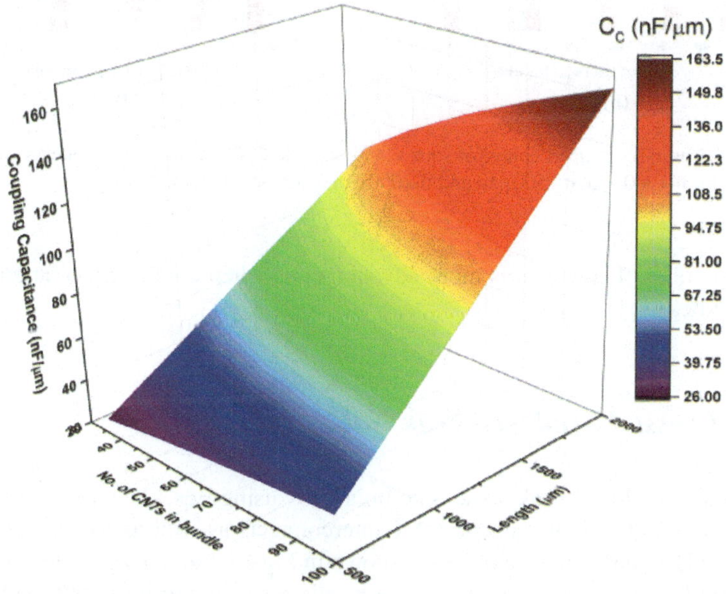

(b)

Fig. 3.25 Coupling capacitance at various lengths and number of CNTs in bundle **a** square bundle, **b** triangle bundle (Sathyakam et al. 2020)

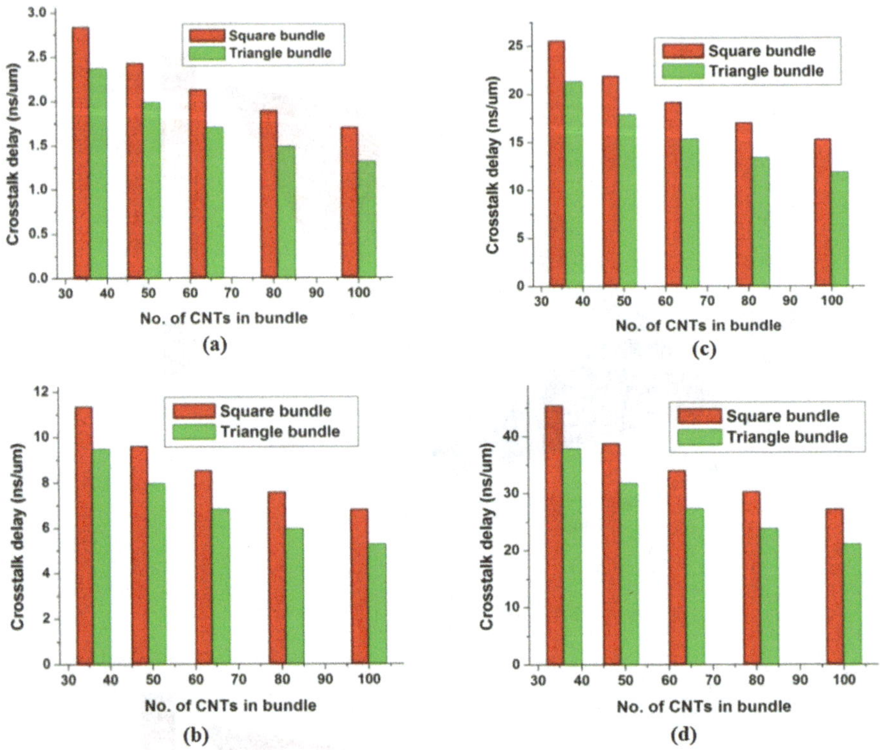

Fig. 3.26 Crosstalk-induced delay for square and triangle CNT bundle interconnects for lengths of **a** 500 μm, **b** 1000 μm, **c** 1500 μm and **d** 2000 μm (Sathyakam et al. 2020)

smaller bundles. This again is more advantageous when compared to larger square CNT bundles.

3.5.5 Crosstalk-Induced Peak Noise

The crosstalk-induced peak noise is computed by using Eqs. (3.1) and (3.2) for both semiconducting CNT-surrounded CNT interconnects as well as for square and triangular CNT bundle interconnects (as shown in Figs. 3.1 and 3.9). Table 3.13 shows the effect of using s-CNTs on the crosstalk-induced peak noise for input voltages of 0.8, 0.5 and 0.3 V. For all the cases of input voltages, the peak crosstalk noise at the output is reduced by 11 to 23%, when the victim is idle. Further, the output noise at the victim side reduced by around 12% when the victim is energized. This shows that using s-CNTs at the periphery is beneficial for CNT interconnects as they act as crosstalk and noise shields.

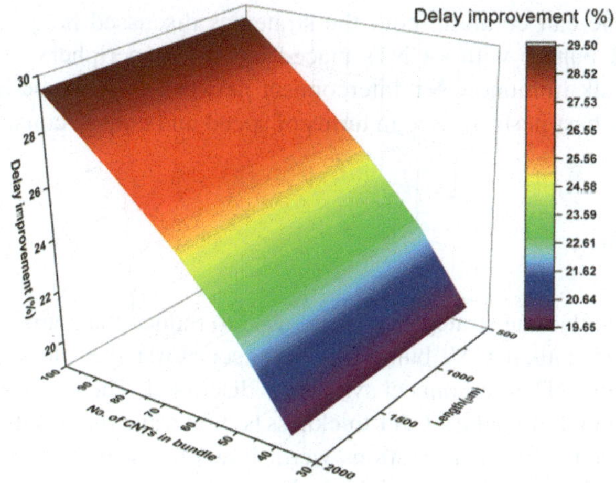

Fig. 3.27 Delay improvement of triangle interconnects over square interconnects for various number of CNTs in a bundle and at various lengths (Sathyakam et al. 2020)

Table 3.13 Crosstalk-induced peak noise voltages where semiconducting CNTs are used at the interconnect periphery

Input voltage (V)	Output Noise (V) when victim is idle, energized	
	(without s-CNTs)	(with s-CNTs)
0.8	0.779, 1.487	0.568, 1.012
0.5	0.439, 0.721	0.377, 0.558
0.3	0.29, 0.621	0.26, 0.602

Table 3.14 Crosstalk-induced peak noise voltages for triangular CNT bundles are used at the interconnects

Input voltage (V)	Output noise (V) when victim is			
	Idle (Square bundles)	Idle (T-CNT bundles)	Energized (Square bundles)	Energized (T-CNT bundles)
0.8	0.779	0.562	1.487	1.012
0.5	0.439	0.373	0.721	0.549
0.3	0.29	0.258	0.621	0.601

Next, the output peak noise of triangular CNT bundles is compared with that of square CNT bundles. Table 3.14 shows the crosstalk-induced noise voltages. It can be seen that T-CNT bundle interconnects are far better than square CNT bundle interconnects in terms of crosstalk-induced peak noise voltage. However, the noise varies around ±0.5% when compared to the s-CNT-covered CNT bundle interconnects.

This means that both the strategies are effectively reducing the crosstalk as well as output noise in CNT interconnects which are beneficial to future chip designers.

Moreover, if one can combine both the strategies discussed here, i.e. by using a triangular CNT bundle with s-CNTs placed around its periphery, one can design a highly crosstalk-immune CNT interconnect network that can be highly reliable which gives the best performance in terms of speed and energy consumption.

3.6 Summary

This chapter detailed the strategies that are used to reduce the coupling capacitance and hence the crosstalk in CNT bundle interconnects. First, the role of s-CNTs as EM wave blockers in CNT interconnect systems is discussed. The modified capacitance expression after including the s-CNT thickness is shown and the coupling capacitance values are derived. SPICE simulations were shown to ascertain the EMI-blocking property of s-CNTs. Interconnect delay values were calculated and compared with and without s-CNTs which reveal that crosstalk and induced delay is reduced when s-CNTs were used at the periphery of the CNT interconnect.

Chapter 4
Ultra-low-k Dielectrics in CNT Interconnects

Silicon-based low-k dielectrics are found to have many reliability problems. Ultra-low-k materials that are non-silicon based are discussed in this chapter. Emerging ultra-low-k materials are polymer-based dielectric materials. They are, namely, graphene oxide/polyimides (GO/PI), boron nitride nanotubes (BNNTs), metal organic frameworks (MOFs) and small diameter CNTs. These materials are found to have very small dielectric constants as low as 1.6. Properties of these materials, like dielectric constant, resistance to oxidation, hydrophobicity, thermal conductivity, Young's modulus, etc., are enlisted and considered for integrating these ultra-low-k dielectrics with CNT interconnects. The challenges associated with integration are also discussed in this chapter.

4.1 Introduction

Interconnects in integrated circuits are separated by insulating dielectric materials that electrically isolate two interconnects. Ideally, the dielectric material should be a perfect insulator that has a very low dielectric constant. However, the low dielectric constants of the currently used dielectric materials range from 2.8 to 3.9 which is not sufficient to meet the demand of ultra-low dielectric constant materials for future interconnects. This problem is mainly due to the high packing densities of integrated circuits which make it difficult to scale down the size of materials. As the distance between adjacent interconnects decreases, there is a need to use ultra-low-k (ULK) dielectrics to overcome crosstalk and electromagnetic interference issues. So this chapter discusses the use of small diameter CNTs as ULK dielectrics for the first time. Further, emerging ULK materials like graphene oxide/polyimides, boron nitride nanotubes and metal organic frameworks are also discussed. Finally, the properties of these emerging materials that are useful for dielectric applications are enlisted.

P. U. Sathyakam and P. S. Mallick, *Design and Crosstalk Analysis in Carbon Nanotube Interconnects*, https://doi.org/10.1007/978-981-15-8888-4_4

Low-k dielectrics are materials that are used as inter-wire barriers to isolate adjacent interconnects thereby reducing crosstalk-induced delay between them. Periodic reviews on fabrication methods (Lionti et al. 2015), reliability challenges (Wu et al. 2015), nanoporous polymers (Ma et al. 2013) and other aspects of low-k materials were presented by many researchers (Farrell et al. 2011; Hatton et al. 2006; Shamiryan et al. 2004). Many issues like porosity, moisture uptake, adhesion, thermal, mechanical and chemical stability of present low-k materials are the causes of concern. Unfortunately, the solution to one limitation leads to another problem in most of the current methods and materials (Baklanov et al. 2015). Further, it must be noted that in ultra-deep nanoscale (UDNS) technologies where future transistors and interconnects will be functioning, new bottom-up approaches will be needed for growing and integrating low-k materials.

This section discusses about emerging ultra-low-k materials for CNT interconnect technologies for the first time. This section is intended to bridge the gap between materials scientists and the IC engineers by discussing the challenges that will be faced by them while integrating ULK dielectrics into CNT interconnects. The physical, chemical, mechanical, electrical and thermal properties of emerging ULK materials are discussed that influence the researchers in deciding the process steps of CNT/ULK interconnect integration.

The dielectric constant of a material depends on its polarizability, which again depends on the induced dipole moment when the dielectric is kept in an electric field. So, the number of polarizable atoms must be reduced to lower the dielectric constant of a material. Also, atoms with small polarizability can be added to reduce the dipole moment. This is because, when a material is subjected to a uniform electric field E, a dipole moment p is induced on the dipoles that are present in it. One can relate E and p using a quantity called polarizability tensor α by $p = \alpha E$. So, materials that have less polarizable bonds (polarizability in Å) like C-C (0.53), C-F (0.56), C-O (0.58), C-H (0.65), O-H (0.71), C=O (1.02), C=C (1.64), CC (2.04) and CN (2.24) are ideal as low-k dielectrics (Shamiryan et al. 2004).

Unlike Cu interconnects, which are fabricated by a top-down approach using dual damascene CMOS process, CNTs are first grown on a substrate and transferred to desired locations by dielectrophoresis followed by densification and chemical mechanical planarization (CMP) (Baklanov et al. 2015). It is very important that in a bundle of CNTs, all the CNTs are connected between the end contacts properly and are aligned along a single direction only (Shulaker et al. 2016). This helps to avoid contact of adjacent CNTs of two different bundles, if any. Additionally, a low-k material must be used to isolate two adjacent CNT bundles so that the RC delay can be kept minimal. The idea of integrating ULK materials in CNT interconnects is still evolving and it demands an extensive research work. There are specific requirements that need to be met for design and integration of ULK materials in CNT bundle interconnects. Materials like organosilicates (SiCOH) have been studied as the low-k dielectrics for copper interconnects (Verdonck et al. 2014); (Gambino 2011). Plasma treatment of SiCOH films was done and after inducing porosity, porous SiCOH films (pSiCOH) were used as interlayer dielectrics (ILD) in copper interconnects (Myers et al. 2015). However, during the integration process, its mechanical strength was decreased thereby reducing its reliability as a low-k interface material. Organic polymers were also studied as an alternative to Si-based dielectrics (Baklanov et al. 2015). They are less polarizable and hence can offer lower dielectric constants (1.2 to 2.5). But, they are mechanically and thermally unstable as they have high coefficient of thermal expansion compared to other materials in an IC. So, they must be chemically treated and processed so that they become suitable for integration as low-k dielectrics. Most of the methods that were used to lower the k-value had failed at the integration stage. For example, porous carbon-doped oxides (p-OSGs) were deposited using plasma-enhanced CVD, which is the most preferred method to deposit films that are compatible with current technology (Vanstreels et al. 2015). The lowest k-value that can be achieved using PECVD is around 2.0. So, alternate methods and materials are needed to replace the existing ones for future interconnect technologies. Some researchers proposed air gaps as the dielectric medium in Cu wires and studied its feasibility and impact (Kumaresan et al. 2014; Zahedmanesh et al. 2015). Current research shows that an effective μ value of up to 2.25, including the barrier, cladding and the dielectric thicknesses, is achievable without any reliability problems for 22 nm technology node (Stan et al. 2015).

4.2 Emerging Ultra-low-k Materials

Alternate materials, approaches and integration techniques are needed to achieve low-k values less than 2 for 14 nm technology nodes and beyond. So, emerging materials like boron nitride nanotubes (BNNTs), graphene oxide (GO) and its composites, metal organic frameworks (MOFs) and small diameter carbon nanotubes are discussed as the prospective candidates of ULK dielectrics. The concept of using small diameter CNTs as prospective ULK dielectrics is proposed here for the first time.

Boron nitride nanostructures are gaining popularity as they have properties similar to carbon nanotubes and graphene. Most of the analytical approaches used to study CNTs and graphene are equally applicable to boron nitride nanostructures. Successful growth and synthesis of carbon nanotubes motivated researchers to contemplate on creating stable boron nitride nanotubes (BNNTs) (Rubio et al. 1994; Blase et al. 1994). BNNTs constitute a 1:1 ratio of B and N atoms in a tubular chain of hexagonal lattice. Basically, BNNTs are found to be semiconducting with a bandgap of 2 eV at 5.6 Å diameter and decreases with diameter irrespective of chirality. Later, the bandgap was found to be 5.5 eV which is independent of tube diameter, chirality and number of tube walls (Chopra et al. 1995; Ohba et al. 2001). Both single-walled and multi-walled BNNTs are synthesized and their growth mechanisms, fabrication and characterization, and their properties were discussed in detail (Ohba et al. 2001; Fuentes et al. 2003; Terauchi 2006; Yap 2009).

BNNTs have excellent physiochemical characteristics like high hydrophobicity, resistance to oxidation and heat, and radiation absorption which make them one of the ideal candidates as ultra-low-k dielectrics (Kalay et al. 2015). The partial

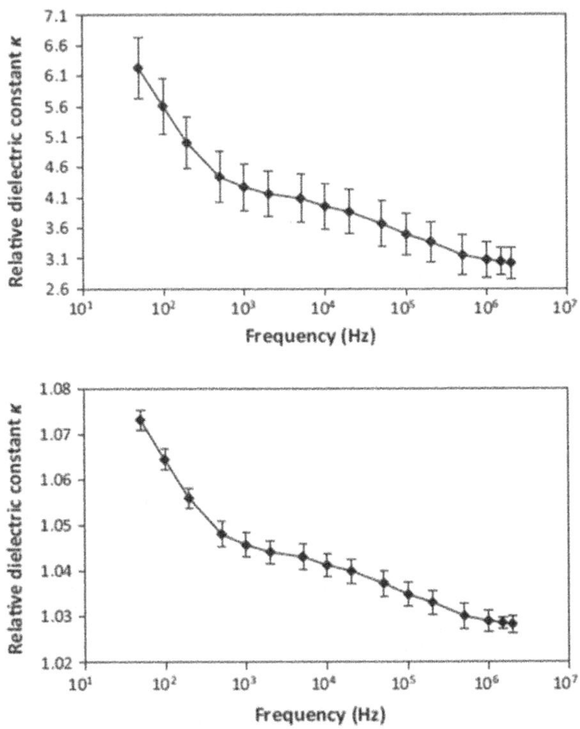

Fig. 4.1 Relative dielectric constant of BNNT solid and BNNT mats at varying frequencies (Hong et al. 2016)

ionic nature of B-N bonds makes BN nanosheets an intrinsic insulator and a large bandgap (5.5 eV) semiconductor. However, its dielectric constant ranged from 4 to 9 depending on the load fraction of poly-methyl-metha-acrylate (PMMA) (Wang et al. 2012).

In a nutshell, BNNTs have very small μ of 1.02 to 1.08 (see Fig. 4.1) (Hong et al. 2010), have high resistance to oxidation at 800 °C (see Fig. 4.2 and are mechanically and thermally strong. So, BNNTs are really promising as ULK dielectrics for future IC technologies, though they have a moderate thermal conductivity ranging from 200 to 450 W/mK axially as shown in Fig. 4.3.

The next emerging ULK material is graphene oxide/polyimides (GO/PI). Graphene is a well-known material for its extraordinary physical and chemical properties which makes it an ideal material to form nanocomposites with other materials. Recently, graphene oxide/polyimide (GO/PI) nanocomposites are studied as ULK dielectrics for nanoelectronic circuits (Wang et al. 2011). Generally, graphene can exist in various forms like nanoribbons (GNRs), nanosheets (GNSs), nanoplatelets (GNPs) and as graphene oxide (GO). These materials have unique properties like high aspect ratio, unique graphitized planar structure and low cost manufacturability. Among these materials, GO has the widest variety of applications. GO is basically a two-dimensional structure that contains covalently bonded carbon atoms which are attached with various oxygen-bearing groups (like hydroxyl, epoxide and carbonyl

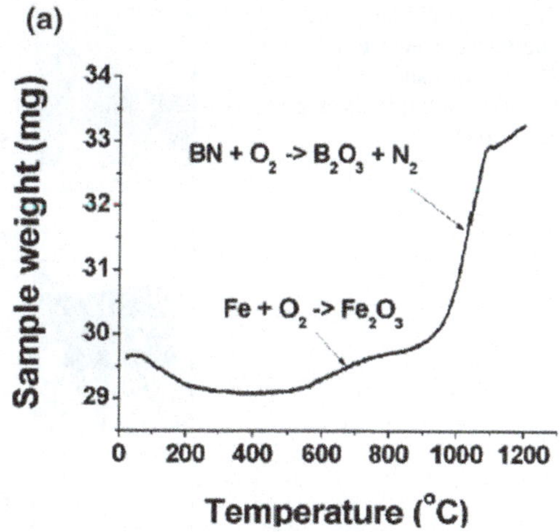

Fig. 4.2 Weight changes of BN sample versus temperature during heating at TGA (Chen et al. 2004)

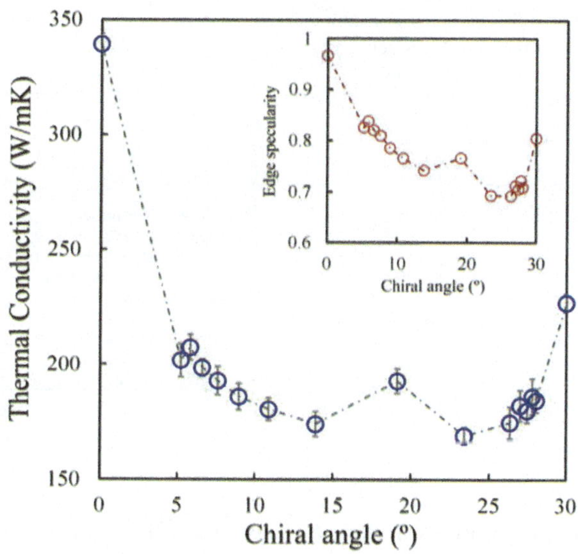

Fig. 4.3 Chiral angle dependence of BNNT thermal conductivity (Kumar and Parashar 2016)

groups). A detailed description of various chemical aspects of graphene oxide is discussed by Gao (2015).

Further, polyimides (PIs) have extraordinary properties of high thermal conductivity, high electrical resistivity, chemical inertness and excellent mechanical properties (Wang et al. 2011). Polyimides were first produced in 1908 by Marston Bogert as discussed in the book chapter by Ohya et al. (1997). But, most of the research started in 1955 and due to their extraordinary heat-resistant properties, polyimides found applications in many fields such as temperature plastics, adhesives, dielectrics and so on (Liaw et al. 2012). An extensive review of syntheses, properties and applications

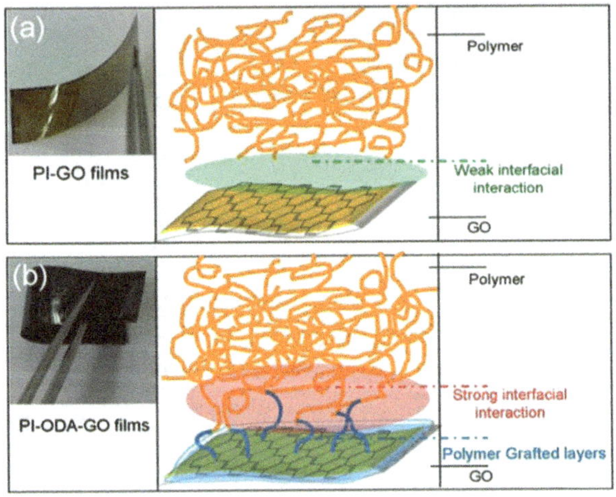

Fig. 4.4 A model of the interphase structure of PI-GO films and PI-ODA-GO films (Wang et al. 2011)

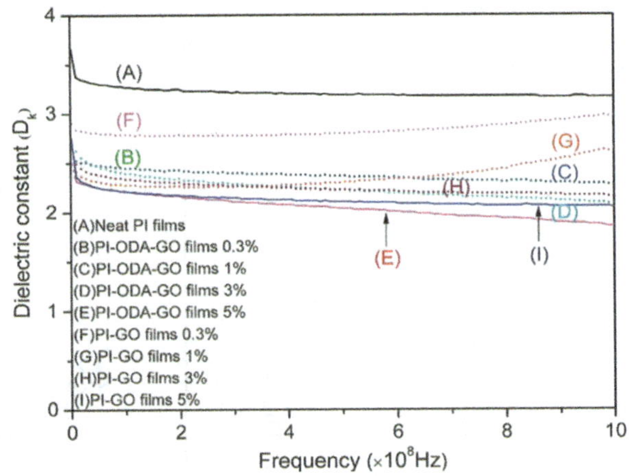

Fig. 4.5 Dielectric constants as a function of frequency of different GO/PI samples (Wang et al. 2011)

of polyimides was done by Liaw et al. (2012). Polyimides itself are not a good choice as low-k dielectrics as they are having poor mechanical stability though they have a dielectric constant of 3.3 that is required for an interlayer dielectric (ILD). However, by adding GO to PI to form a GO/PI composite, the dielectric constant was decreased to around 2.0 for 5 wt% concentration of GO in PI. A similar reduction was experienced for a 5 wt% concentration of PI-ODA-GO films. The presence of GO and 4,4-diaminodiphenyl ether (ODA) in the PI films increases its electrical resistivity and a dielectric constant of around 2.0 was measured (Wang et al. 2011) (see Figs. 4.4 and 4.5).

GO/PI composites can be grown using a solution blending method where GO was mixed with a suitable base agent like dimethylacetamide (DMAc). Initially, GO was formed by stirring of natural graphite (produced by chemical vapour deposition method), potassium permanganate, sulphuric acid and sodium nitrate. Remarkably,

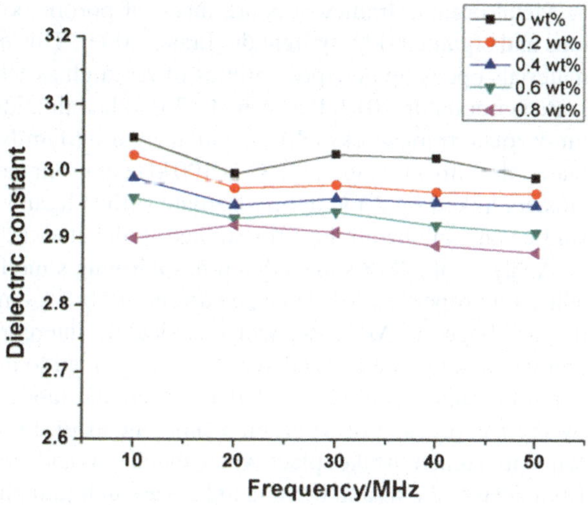

Fig. 4.6 Dielectric constants as a function of wt% of PI (Bai et al. 2016)

they have very high tensile strength of 844 MPa and Young's modulus of 20.5 GPa which is ideally suited for low-k dielectric applications. The presence of graphene makes the composite mechanically and thermally strong. Integration of GO-NH_2 (4,4-diaminodiphenyl ether, ODA-GO) into polymer mats was achieved. One of the important factors that determines the reliability of ODA-GO is the polymer interaction that binds the GO and the polymer together in a mat form (Fig. 4.4).

Strong interfacial interaction was found between GO and PI-ODA polymers which is necessary to bind the material to its base. After various steps of preparation and fabrication, PI-ODA-GO films were found to have considerably lower dielectric constant of around 2.0 which decreases with increase in the weight fraction of ODA-GO in the sample. Another work shows that incorporation of only 5 wt % of GO in the PI resulted in 12X and 18X increase in tensile strength and tensile modulus of the GO/PI composite (Ha et al. 2012). Further, reduced GO (rGO), formed by Hummers method, showed relatively lower tensile properties than GO. Better performance of GO was attributed to good dispersion and effective stress transfer between the GO and PI, and hence better mechanical properties of GO/PI composites.

The thermal conductivity of GO composites is also a major factor that implies the stability of GO polymers at high operating temperatures of integrated circuits. Tseng et al. (2013) had studied the thermal and dimensional stabilities of GO/PI composites. They had first synthesized glycidyl methacrylate-grafted graphene oxide (g-GO) into the PI-matrix. The resulting polymer is a PI/g-GO nanocomposite that exhibited increase in thermal conductivity to 0.81 W/mK from 0.13 W/mK when g-GO weight fraction was less than 10 wt%. Graphene/PI composites were found to have good moisture barrier properties making them ideal as low-k materials (Tsai et al. 2013; Yoo et al. 2014). However, the work was done on reduced graphene. So, there is a need to find out the moisture uptake property of graphene oxide/PI composites as they show promising results as future low-k materials.

Metal organic frameworks are made of porous polymeric chains linking metal ions and organic bridging ligands (James 2003). This new class of crystalline hybrid materials opens up new possibilities of research as ULK dielectrics in UDNS integrated circuits. In 2012, Redel et al. (2013) had first identified single crystal zeolitic imidazolate frameworks (ZIFs), which are a subfamily of MOFs, as potential low-k materials with a k-value of 2.33 at 100 kHz at room temperature. ZIF-8 (zinc(II) ion clusters linked by 2-methylimidazolate (mIm) ligands [Zn(mIm)2]) were the most studied one among all the ZIFs studied Redel et al. (2013).

Analysis of ZIF-8 shows that it has a Young's modulus greater than 3GPa, good adhesion properties, low leakage current of 10-8 A cm2 at 1 MV cm-1 and a breakdown voltage of 2 MV cm-1 which is ideal for interconnect/low-k applications. Further details of processes and synthesis carried out to prepare ZIFs were discussed in detail by Zagorodniy et al. (2010). Systematic fabrication of MOF involves layer-by-layer deposition of MOF on a substrate as discussed in Stavila et al. (2014). A sequential reaction takes place where metal ions and the organic ligands bind together. Both the metal ions and organic linkers are in liquid phase. Gradually, the metal ions bind to the upper layers of the ligands. In situ surface plasmon studies show that the binding process is linear where the MOF film thickness is proportional to the number of immersion cycles. Interaction of MOFs with the substrate surface and their orientation is a major factor to be considered during low-k integration. Dielectric properties of a class of $Zn_4O(CO_2)_6$-based MOFs were theoretically analysed using the Clausius–Mossotti relation (Zagorodniy et al. 2010). Static dielectric constants of a set of 30 MOFs derived from the C-M equation show that the dielectric constant ranges from 1.21 to 1.94. However, only three MOFs showed good elastic and bandgap properties desirable for low-k applications in ICs. They are IRMOF-11 (k = 1.45, bulk Young's modulus [YM] = 12.6 GPA, Eg = 4.91 eV), (COOH)2 (fcc) (k = 1.62, bulk YM = 24.3 GPa, Eg = 4.17 eV) and (COOH)2 (sc) (k = 1.62, bulk YM = 22.3GPa, Eg = 5.56 eV). A well-detailed device fabrication methodology of MOF thin films for microelectronics was discussed by Allendorf et al. (2011). Controlled deposition of MOF films on a substrate that are dense, homogeneous and robust was done. There are generally three ways in which this can be done, namely, in situ, ex situ and seeding methods. A detailed understanding of these methods can be found in Allendorf et al. (2011). Ex situ synthesis involves auto-assembly of preformed objects like nanocrystals and colloids. In situ growth method is more suitable for device integration like low-k dielectrics in microelectronic chips. So, high precision patterning and growth control must be achieved here. Seeding mechanism is more appropriate for gas sensor applications and not for device fabrication and integration. This is because, currently, there are no masking and patterning techniques developed for integrating seeded MOFs with devices. Further, various functionalization mechanisms like ionic, covalent and metal-ion exchange were discussed by Allendorf et al. (2011). Device integration, manufacturing issues and fabrication challenges are also well covered. MOFs are found to be highly thermally stable till 400 °C, and have good mechanical and thermal stabilities Mendiratta et al. (2015). The simulated dielectric constant of IRMOF-M2c [Zn4O(tdca)3] (tdca = tetracene-2, 8-dicaboxylate) is

Table 4.1 Properties of emerging ultra-low-k materials

Material	Property	Value	References
BNNTs	Dielectric constant (matted BNNTs with air voids)	1.02–1.08	Hong et al. (2016)
BNNTs	Bandgap	5 eV	Blase et al. (1994)
BNNTs	Resistance to oxidation in air	Up to 1000 °C; 800 °C	Yap (2009), Chen et al. (2004)
BNNTs	Hydrophobicity	Contact angle up to 170°	Kalay et al. (2015)
BNNTs	Bulk modulus	26 GPa	Ohba et al. (2001)
BNNTs	Axial Young's modulus	0.5–0.6 TPA	Ohba et al. (2001)
BNNTs	Thermal conductivity (SW BNNTs)	Axial-200–350 W/mK; 120–960 W/mK	Kalay et al. (2015)
BNNTs	Thermal conductivity (MW BNNTs)	350–450 W/mK	Kalay et al. (2015)
BNNTs	Thermal conductivity (Transverse/radial)	20 30 W/mK	Terao et al. (2010)
Graphene oxide and its composites			
PI-ODA-GO, PI-ODA-GO films 5%	Dielectric constant	2.0	Wang et al. (2011)
PI/rGOD (5 wt%); PI-ODA-GO films 5%	Tensile strength (MPa)	1186; 728	Wang et al. (2011), Ohya et al. (1997)
PI/rGOD (5 wt%); PI-ODA-GO films 5%	Young's modulus (GPa)	26.19; 19.6	Wang et al. (2011), Ohya et al. (1997)
PI/rGOD (5 wt%)	% Elongation	8.66	Ohya et al. (1997)
ODA-GO, PI/GO and PI/rGO; PEDOT-PSS/WPPy/PVA/GO	Thermal degradation	200 °C; 220 °C 310 °C; 200 °C	Wang et al. (2011), Ohya et al. (1997), Joshi and Deshmukh (2016)
Pi/g-GO	Thermal conductivity	0.81 W/mK	Liaw et al. (2012)

1.21. Factors that influence the dielectric constant such as anions, solvents and organic linkers were also studied. Table 4.1 enlists all the properties of MOFs that make them a prospective candidate as ULK dielectric materials. The class of MOFs and their dielectric constants are summarized. MOFs that have μ in the range of 1.94– 2.4 make them a prospective candidate as ULK dielectrics for CNT interconnects. Their small pore size of 3.4 Å, very small leakage current of 10-8 A/cm^2 and good stability up to 350 °C are ideal for CNT interconnect applications. However, their very low thermal conductivity of 1.5 W/mK is a cause of concern and needs to be addressed. Latest advances in synthesis of novel MOFs and their dielectric characteristics show that MOFs are indeed a prospective candidate for low-k dielectric applications (Chen

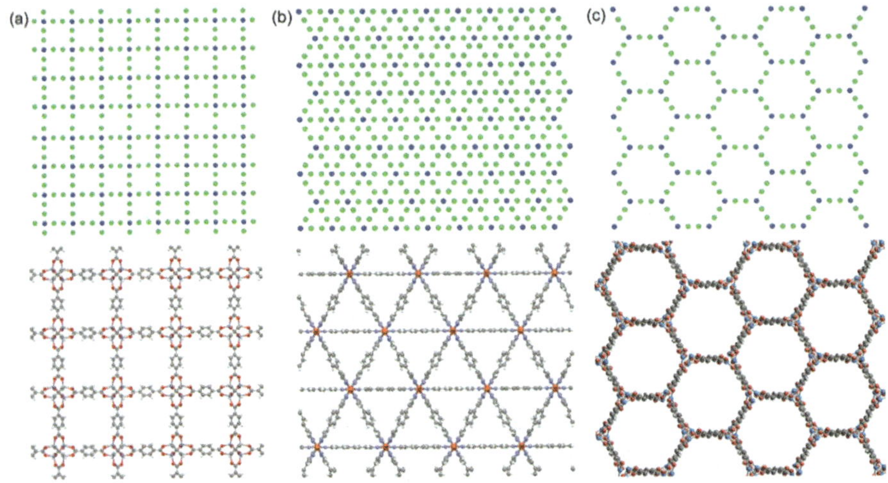

Fig. 4.7 Cross-sectional views of **a** the idealized simple cubic structure (top) based on the real material IRMOF-1 (bottom), **b** the idealized triangular-channel structure (top) based on the real material FeBDP (bottom), and **c** the idealized hexagonal-channel structure (top) based on the real material MOF-74 (Babaei et al. 2017)

et al. 2016; Dhieb et al. 2015; Guo et al. 2016; Li et al. 2016; García-Fernández et al. 2016; Sieradzki et al. 2016; Abhyankar et al. 2016; Wang et al. 2016). Temperature dependence of the dielectric constants of MOFs was studied. All the results reveal that the MOFs are highly stable up to around 420 K, which means that MOFs are highly stable at high-temperature environments in tightly packed ICs. The effect of pore shape and size of MOFs on its thermal conductivity was investigated by Babaei et al. (2017). Various geometries of pore channels like simple cubic, triangular and hexagonal channels were studied as depicted in Fig. 4.7. It was found that MOFs having larger pores exhibited lower thermal conductivity than smaller pores (see Fig. 4.8a). MOFs having hexagonal pore channels showed highest and lowest thermal conductivities along its axial and orthogonal directions, respectively, as shown in Fig. 4.8b.

4.3 Comparison of Ultra-low-k Dielectric Materials

Table 4.1 and 4.2 gives the list of properties of the emerging ultra-low-k materials. These properties are helpful in understanding their fabrication and integration processes for low-k dielectric applications.

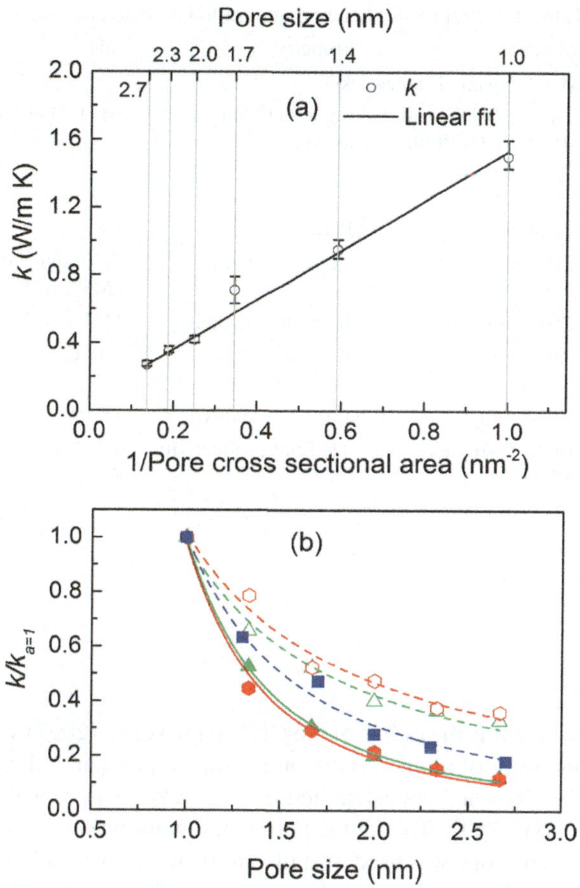

Fig. 4.8 **a** Thermal conductivity of the simple cubic structure without adsorbed gas as a function of pore size (top axis) and crystal density (bottom axis). **b** thermal conductivity of MOF crystal's triangular (triangles) and hexagonal (hexagons) channels parallel (filled symbols) and perpendicular (empty symbols) to the channel directions along with simple cubic structure (filled squares) as a function of pore size (Babaei et al. 2017)

4.4 Small Diameter CNTs as ULK Dielectrics

Many researchers studied dielectric properties of CNTs (Benedict et al. 1995; Torrens 2004; Kozinsky and Marzari 2006; Lu et al. 2007; Fagan et al. 2007; Zhao et al. 2004; Guan et al. 2008). The first work on the analysis of static polarizabilities of SWCNTs was done by Benedict et al. (1995). They found that the radial polarizability, α_{xx}, of single-walled CNTs is strongly dependent on the radius of the CNTs and is proportional to R^2. But, α_{xx} is independent on the nanotube chirality [100]. So, like boron nitride nanotubes, CNTs with small diameters can have very low dielectric constants. The relation between the radial polarizability α_{xx} and the radial dielectric constant $\epsilon_\perp$, considering CNT as a perfect cylinder, can be given as (Lu et al. 2007)

$$\alpha_{xx} = \frac{1}{2}\frac{\epsilon_\perp - 1}{\epsilon_\perp + 1}R^2 \tag{4.1}$$

Table 4.2 Properties of emerging ultra-low-k materials—continued

Material	Property	Value	References
Metal Organic Frameworks			
Class of MOFs and IRMOFs ZIF-8 film	Static dielectric constant	2.4; 1.21–1.94, 2.33	Usman et al. (2015), Zagorodniy et al. (2010), Mendiratta et al. (2015)
ZIF-8 film	Pore size	3.4	Redel et al. (2013)
ZIF-8 film	Leakage current	10^{-8} Acm^2 at 1 MV cm^{-1}	Redel et al. (2013)
ZIF-8 film	Breakdown voltage	2 MV cm^{-1}	Redel et al. (2013)
ZIF-8 film	Resistance to oxidation	350 °C	Redel et al. (2013)
ZIF-8 film	Elastic modulus	3 GPA	Redel et al. (2013)
MOF-5, single crystal MOF	Thermal conductivity	0.32 W/m K, 1.5 W/m K	Huang, McGaughey and Kaviany (2007), Huang, Ni, Millward, McGaughey, Uher, Kaviany and Yaghi (2007), Babaei et al. (2017)

where R is the radius of the CNT. Fagan et al. (2007) showed that the radial dielectric constant of aligned semiconducting carbon nanotubes (s-CNTs) ranges from 1.6 to 1.7. Researchers calculated α_{xx} for MWCNTs as well as 2D and 3D arrays/bundles of SWCNTs. The first experimental work was done in 2007 where a scanning force microscope was used to apply an AC field across a CNT to find its dielectric constant as depicted in Fig. 4.9 (Fagan et al. 2007). Moreover, Kozinsky and Marzari (2006) had done the polarizability calculations for MWCNTs and bundles of SWCNTs. Kozinsky and Marzari (2006) and Lu et al. (2007) confirm the model given by Eq. (4.1) and it is concluded that the radial dielectric constant of a CNT is independent of its chirality. After performing DFT-based calculations, which take into account the local field effects, which are neglected in tight binding calculations in Benedict et al. (1995), the polarizability of a bundle of CNTs is given by

$$\alpha_{\perp}^{b} = \frac{\Omega}{4\pi}(\epsilon_{\perp} - 1) \tag{4.2}$$

where $\Omega = L^2$ is the cross-sectional area of a unit cell. Again, the single nanotube polarizability can be related with bulk polarizability as

$$\alpha_{perp} = \frac{\Omega}{2\pi}\frac{\epsilon_{\perp} - 1}{\epsilon_{\perp} + 1} = \frac{\alpha_{\perp}^{b}}{1 + \frac{2\pi}{\Omega}\alpha_{\perp}^{b}} \tag{4.3}$$

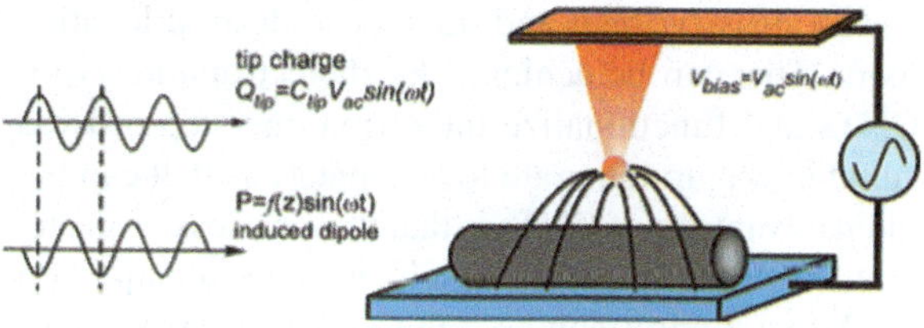

Fig. 4.9 Schematic illustration of the scanning force microscopy approach to the near-dc polarization of individual carbon nanotubes (Fagan et al. 2007)

Growing of small diameter CNTs is the first step towards achieving low dielectric constants. Some works were done by Zhao et al. (2004) who achieved a diameter of 3. Another work was reported by Guan et al. (2008) where they have shown CNTs of (3,3), (4,3) and (5,1) chiral indices with diameters ranging from 1.0 to 1.2 nm. However, dielectric constants were calculated using (4.1) for CNTs up to 1.4 nm diameter only (Lu et al. 2007). The validity of (4.1) for CNTs of diameter less than 3 nm is not known. So, a new model must be formulated relating the polarizability to the dielectric constant for CNTs of diameter less than 1.4 nm. Further, more work needs to be done on growing small diameter CNTs into mats and characterizing its dielectric properties for integrating them in interconnects. The main factors that contribute towards the dielectric constant of CNT mats are the diameter of CNTs, orientation and distribution of CNTs and the inter-tube spacing which can be thought as air gaps. As the inter-tube spacing increases, the number of tubes, and hence number of dipoles decreases. Consequently, its dielectric constant decreases. While lot of work concentrate on investigating the dielectric properties of large diameter CNTs and CNT composites (Amrin and Deshpande 2016; Yang et al. 2016; Li and Lue 2007; Iqbal et al. 2015; Poh et al. 2016), not much progress was made in studying the dielectric properties of specific CNTs that have small diameters up to 2nm. So, for low-k dielectric applications, small diameter CNTs must be studied in detail and their growth mechanisms should be understood.

4.5 Integration Challenges

Growth processes and integration mechanisms like etching, patterning, masking and plasma processing that were adopted for traditional Cu interconnects lead to the degradation of the dielectric properties of low-k materials (Baklanov et al. 2013). So, many challenges were faced by researchers and engineers in the microelectronics industry in realizing low-k dielectrics in between Cu interconnects (Baklanov et al. 2015). Integration of low-k dielectrics for CNT interconnects is a new and emerging area altogether when compared to the strides made in copper/low-k integration (Theis and Wong 2017). The foremost challenge in this direction is to precisely grow aligned metallic CNTs horizontally as bundled interconnect wires. Recent advances on horizontal CNT growth and alignment are discussed by Cole et al. (2016). Generally, two methods of low-k integration are possible for CNTs. One way is to grow

CNTs, densify them and transfer to desired location for decoration of low-k polymers. This can be achieved by dielectrophoresis method. Another way is to grow CNTs and functionalize them in a one-step process in a CVD chamber. This is a suitable way to integrate low-k materials that can be grown in a CVD chamber. The main advantage of CNTs is that they are less prone to boundary anomalies and deformations, when grown precisely, which is a major concern in Cu interconnects. Also, CNT/low-k wires can be stand-alone structures that do not require additional barriers, cladding and etch stop layers that are integral part of Cu interconnects. Chemical reaction-based polymer growth can be achieved for materials like GO/PI, GO/ODA and related polymers.

Interaction of the low-k material with CNT sidewalls is a major area of concern as the thermomechanical properties of the low-k material affect the performance of CNT interconnects. Free flow of heat between interconnects and the adjoining low-k materials is necessary to ensure that the dielectric material does not become a thermal bottleneck. Another concern is orientation of low-k materials with respect to the CNT bundle sidewalls which affect the thermal transport. To attain well-defined orientation, the effects of dielectric constant along axial and perpendicular directions of the material need to be studied. Sidewall functionalization of CNTs with non-polar bonds is an emerging field of research (Gebhardt et al. 2011). This aspect can be used to decorate CNTs with low-k materials by chemical reactions. The effect of pore sizes and pore channel shapes on the dielectric and mechanical properties of low-k dielectrics is also seen to contribute towards proper selection of low-k materials with selective pore sizes (Babaei et al. 2017). As discussed in the previous section, improvement in adhesion to surfaces and CNT sidewalls can be achieved as in the case of GO/PI and ODA-GO polymers. For BNNTs, control of chirality can improve the adhesion of the materials to the substrate as well as CNT sidewalls.

Nanoscale buckling is an important phenomenon that occurs in growth of low-k dielectrics due to thermal stress during growth phase. Time-dependent dielectric breakdown (TDDB) tests of emerging low-k materials outlined here must be done to ensure their long-term reliability and performance. A detailed outline of TDDB testing procedures and the results of existing low-k materials are reviewed by Croes et al. (2015).

In the case of metal organic frameworks, a recent study was done on two types of surface anchored MOFs (SURMOFs), namely, HKUST-1 and 2D Cu-BDC (Redel et al. 2013). Their frequency-dependent refractive index is found to increase with frequency. At the same time, the frequency-dependent dielectric constant also increases with frequency as shown in Fig. 4.10. Further, there is a correlation between the dielectric constant and the refractive index of 2d Cu-BDC SURMOF-2 structure which is shown in Fig. 4.11. These two quantities were found to decrease with increase in length of organic linkers. So, for low-k applications, the MOFs should have short lengths of organic linkers.

Fig. 4.10 Dielectric constants of **a** HKUST-1, **b** 2D Cu-BDC at different temperatures and at varying wavelengths (Redel et al. 2013)

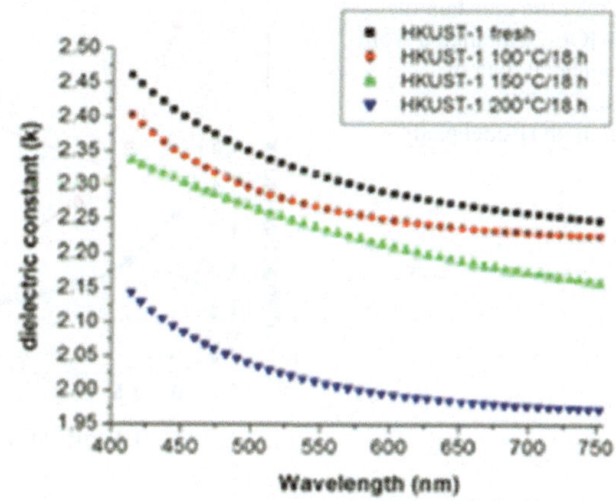

(a)

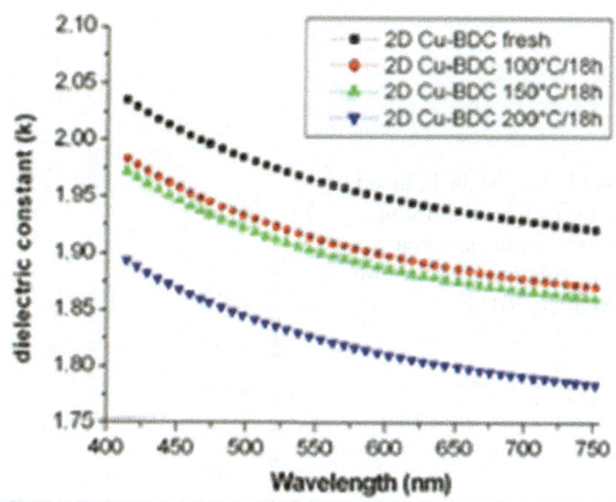

(b)

4.6 Results and Discussions

The radial dielectric constants of small diameter s-CNTs are computed for (n, m) indices of (2,2), (3,3), (4,3) and (5,1) which are experimentally grown and separated till date (Zhao et al. 2004; Guan et al. 2008; Liu, Liu, Li, Bhola, Jackson, Scott, Page, Irle, Morokuma and Zhou 2014) and are listed in Table 4.3.

Figure 4.12 shows the crosstalk-induced delay which decreases with increase in the coupling capacitance when s-CNTs are used. However, it was observed that the

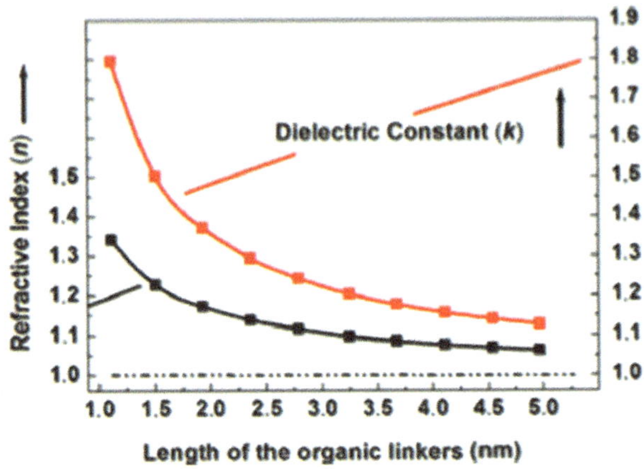

Fig. 4.11 Refractive index and dielectric constant of SURMOF-2 calculated from experimental data of Cu-BDC at 750 nm wavelength (Redel et al. 2013)

Table 4.3 Radial dielectric constants of various s-CNTs

s-CNT type (n, m)	Radial dielectric constant ($\epsilon_\perp$)
(2,2)	2.82
(3,3)	3.62
(4,3)	4.15
(5,1)	4.78

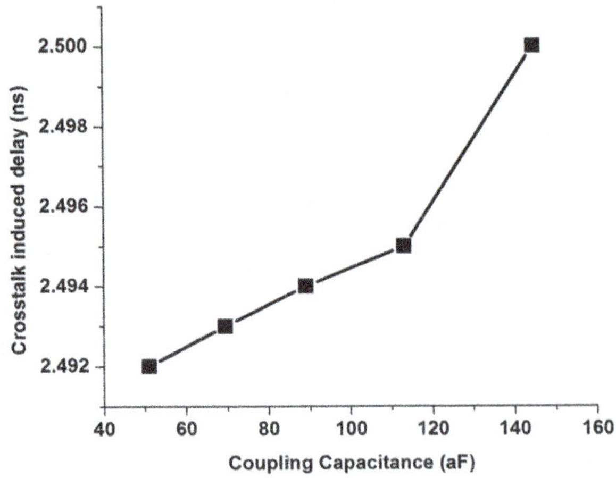

Fig. 4.12 Crosstalk-induced delay for different coupling capacitance values, when s-CNTs are used (Sathyakam and Mallick 2018)

crosstalk-induced delay is 10 times higher if s-CNTs are not used as EMI shields. The interconnects are driven by 14 nm technology node FinFETs from PTM libraries. The proposed model is effective in reducing crosstalk-induced delay compared to conventional CNT bundle interconnects. Further, it is very attractive as the design and fabrication complexities can be reduced drastically and hence costs should be lesser to bring the technology to market.

Chapter 5
Reducing Crosstalk Using Air Gaps

Air is having a dielectric constant of unity and is an ideal dielectric medium in ICs. However, there are many problems associated with air gaps in ICs, especially thermal issues. As trapped air gets hotter, the reliability of interconnects can degrade. So, this chapter explains the performance and reliability of CNT/air-gap interconnects. Performance metrics like RC delay, crosstalk delay, power dissipation and area are analysed.

5.1 Introduction

Optimum performance of CNTs can be achieved if air gaps are used as low-k medium for CNT interconnects. CNTs can be placed on Si substrate precisely by dielectrophoresis. This is based on the recent advancements in realizing MWCNT interconnects in a 1 GHz microprocessor which shows that metallic CNTs can be dispersed at selected locations on silicon by dielectrophoresis (Close et al. 2008). In another work, CNTs were suspended on trenches of Si and their thermal analysis was done (Pop et al. 2006). Unlike Cu/AG interconnects, CNT/AG interconnects do not need additional barriers, cladding and etch stop layers. Rather, one can have multiple CNT interconnects on a single SiO_2 substrate with air as the dielectric medium as depicted in Fig. 5.1. Each interconnect is a planar combination of 3–7 CNTs that form a sheet. Each CNT is having a diameter of 2 nm. The adjacent wires are separated by air gaps of 7–16 nm width (see Fig. 5.2). CNTs on SiO_2 are better option than suspended CNTs as the current flow will be drastically reduced in the later due to self-heating and phonon scattering (Pop et al. 2006; Mohsin et al. 2014).

P. U. Sathyakam and P. S. Mallick, *Design and Crosstalk Analysis in Carbon Nanotube Interconnects*, https://doi.org/10.1007/978-981-15-8888-4_5

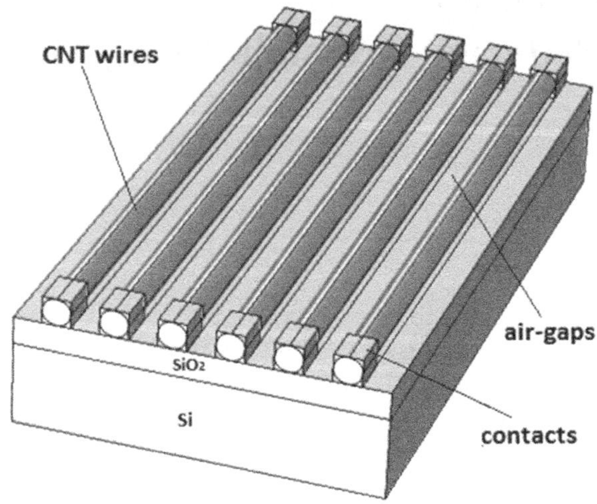

Fig. 5.1 CNT/air-gap structures on SiO_2 substrate (Sathyakam and Mallick 2017)

5.2 Crosstalk in CNT/AG Interconnects

The intrinsic delay of an RC-limited interconnect can be given as

$$\tau = \rho\epsilon\frac{L^2}{Ht} \tag{5.1}$$

where the resistivity ρ and interconnect length L can be reduced by using CNTs instead of copper, the insulator permittivity ϵ can be reduced by using air gap and the interconnect height H can be increased by using arrays or bundles of CNTs. Most importantly, the insulator thickness t should be increased to reduce the delay and the mechanical stability of the structure. However, by using air gaps as the medium, it is possible to decrease the inter-wire distance where d can be kept minimum which can reduce the chip area. In the case of interlayer dielectrics, the distance increases when low-k dielectrics are used. But, as this expansion is in vertical direction, scaling down in lateral direction is unaffected. The coupling capacitance between two corresponding CNTs in coupled CNT bundles can be expressed as

$$C_C = \frac{\pi\epsilon}{cosh^{-1}(y/d)} \tag{5.2}$$

where y is the distance between the corresponding CNTs in interconnects and d is the diameter of each CNT in the bundle. For N number of outer CNTs that are coupled to each other, the coupling capacitance is $N * C_C$. From Fig. 3.3, it can be seen that the capacitance is decreasing with the technology nodes. This is because, as the CNT bundle dimensions are scaled down, the number of CNTs in the bundle decreases and hence the capacitance decreases. Also, the surface area of the CNT bundle decreases

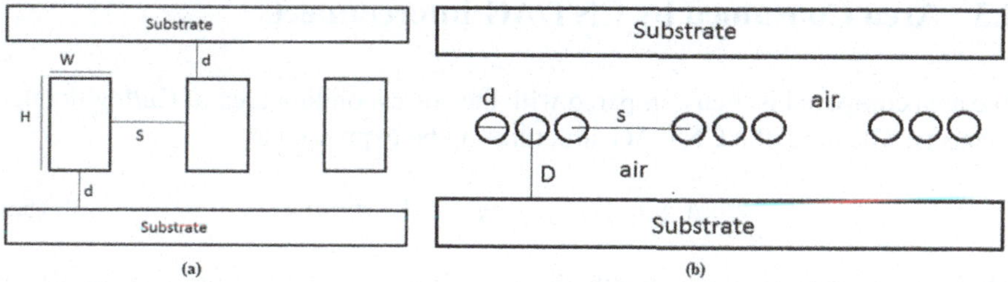

Fig. 5.2 Schematic of **a** copper wires, **b** CNT bundles, considered in this work (Sathyakam and Mallick 2017)

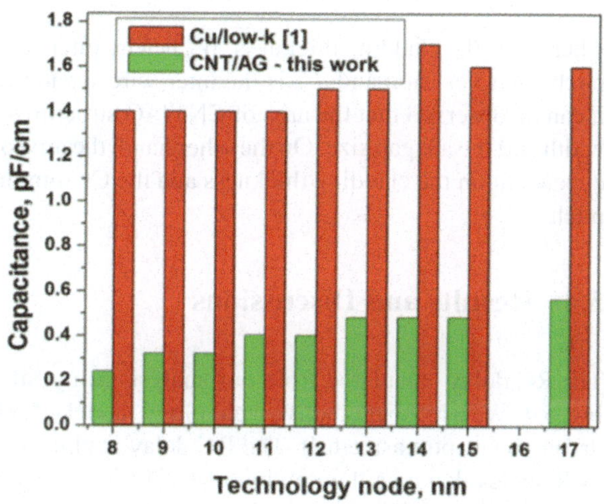

Fig. 5.3 CNT/AG and Cu/low-k capacitances at various technology nodes (Sathyakam and Mallick 2017)

with technology node which contributes to the reduction in capacitance. Further, the resistance of the CNT bundle is dependent on the number of CNTs in the bundle as expressed. So, as the interconnect pitch decreases with technology node, its resistance increases. Thus, the RC delay is dependent on increasing resistance and decreasing capacitance. So, the RC product remains constant across all the technology nodes. The CNT/AG capacitance per centimetre is at least 70% lesser than Cu/low-k values at each corresponding technology node. The ϵ value in ITRS is 4.2 considering the inter-wire dielectric and the diffusion barrier mediums. It must be seen that the L^2/Hd ratio in 'Eq. 5.1' is almost constant across technology nodes (Ceyhan and Naeemi 2013) and is visible in the RC delay analysis in the following section. Here, the type of copper wires considered is shown in Fig. 5.2a and the CNT wires are shown in Fig. 5.2b. It is assumed that the interconnect width is same as the corresponding technology node size. Then minimum wire spacing is used, which is possible only because of air gaps, in computing the area that would be consumed by a CNT/AG interconnect. The capacitance calculated for this structure is shown in Fig. 5.3.

5.3 Area Consumed by CNT/AG Interconnects

The area computed is then compared with the corresponding area of Cu/low-k interconnects. The area of a CNT/AG structure can be expressed as

$$Area_{CNT/AG} = (d_w + d_{ya})^2 nm^2 \quad (5.3)$$

where dw is the interconnect width, dya is the inter-wire separation, same as air-gap size. Also, the area of Cu/low-k interconnect can be expressed as

$$Area_{Cu/low-k} = (W_p + d_c + t_c + d_{yk})^2 nm^2 \quad (5.4)$$

where d_c is the cladding thickness, W_p is wire pitch, t_c is the Cu thinning at minimum pitch due to erosion and d_{yk} is the inter-wire separation. From the above equations, it can be observed that the area of CNT/AG structures is dependent only on the wire width and the air-gap size. On the other hand, the area of Cu/low-k structure is further dependent on the cladding thickness and the Cu thinning due to erosion at minimum pitch.

5.4 Results and Discussions

The RC delay calculated from our analysis using Silvaco SmartSPICE is shown in Fig. 5.4. We use the predictive technology model (PTM)-based 7 nm node inverter as driver (www.ptm.asu.edu/). The RC delay is almost same for all technology nodes as R scales down as the number of CNTs increases in the bundle and C scales up correspondingly. We use an array of 2 nm diameter SWCNTs instead of a rectangular bundle as shown in Fig. 5.4 inset. Table 5.1 shows the technology node, CNT

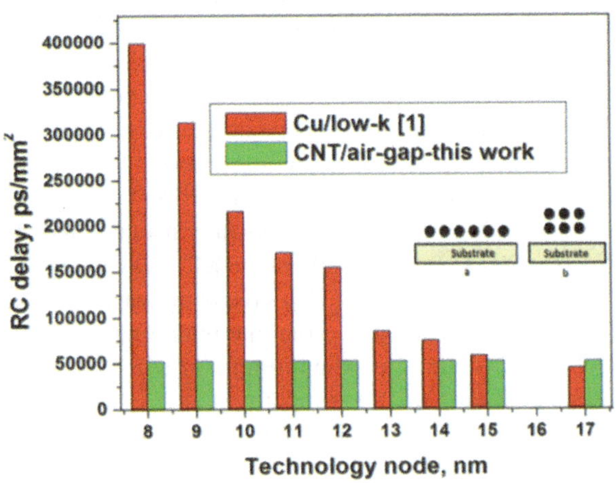

Fig. 5.4 RC delay of CNT/AG wires compared with Cu/low-k at T = 300 K. Inset: geometry of bundle used in this work is type (**a**) (Sathyakam and Mallick 2017)

Table 5.1 Dimensions and area of CNT/AG interconnects compared with Cu/low-k interconnects as per ITRS

Technology node (nm)	W (nm) (N_{CNT})	y (nm)	Area CNT/AG (nm^2) this work	Area Cu/low-k (nm^2)-ITRS	Reduction (%)
8	8 (3)	7	225	1115.56	79.83
9	9 (4)	8	289	1406.25	79.44
10	10 (4)	9	361	1730.56	79.13
11	11 (5)	10	441	2275.29	80.61
12	12 (5)	11	529	2683.24	80.28
13	13 (6)	12	625	3352.41	81.35
14	14 (6)	13	729	4356	83.26
15	15 (6)	14	841	6099.61	86.21
17	17 (7)	16	1089	6756.84	83.88

bundle width W in nm, the number of SWCNTs in a bundle, N_{CNT} and the distance between adjacent wires (air-gap size) y in nm. The area of Cu/low-k interconnects was calculated using the parameters from ITRS and by substituting them in 'Eq. 5.4'. By comparing the area of Cu/low-k and CNT/AG interconnects, it was found that 79%–86% of reduction can be achieved across the technology nodes. In the case of CNT/AG wires, cladding is not needed as they can be dispersed on SiO_2 directly. Also, the wire width of Cu/low-k is more than double the technology node size. Further, the dielectric thickness of low-k materials for Cu wires is higher when compared to the wire spacing that was taken for CNT/AG wires. These are the main reasons behind the reduction of area consumed by CNT/AG wires when compared to Cu/low-k wires. The change in area for CNT/AG wires is within ±4.5% with respect to Cu/low-k wires if the dimensions for minimal RC delay are used. In this way, high-density ICs that can have scaled-down components can be achieved, paving the way for advanced multi-core 3D integrated chips.

5.5 Summary

Carbon nanotube-based interconnects with air gaps as dielectric medium are proposed in this chapter. The coupling capacitance of CNT/AG and Cu/low-k interconnects are calculated and compared. It is found that the capacitance is at least 60% ± 2% lesser than Cu/low-k interconnects. Further, the RC propagation delay is found to be almost constant across different technology nodes compared to Cu/low-k wires. This is due to the scaling down of the interconnects which increases the resistance, while the capacitance is reduced. The area consumed by CNT/AG structures in an IC was found to be 79%–86% lesser than Cu/low-k interconnects. It must be noted that the area varies ±4.5% when the dimensions optimized for lesser R(T)C delay (as discussed in next chapter) are considered.

Chapter 6
Thermal Analysis of CNT Interconnects

In this chapter, the effects of temperature are considered by taking into account the temperature coefficient of resistance R_T of CNTs. The R_TC delay is found out to be highly temperature dependent. Then, CNTs are treated as fractals and the fractal dimensions of metallic CNTs are modelled. This fractal dimension is used to formulate the fractional differential equations of fractal order to solve the heat flow in CNTs.

6.1 Thermal Analysis of CNT/Air-Gap Interconnects

Thermal, mechanical and chemical stabilities of low-k dielectrics are the major concerns of present-day IC designers. Even though many researchers are focusing on developing ultra-low-k dielectrics for next-generation interconnects, they have not yet achieved the all-round stability (Shamiryan et al. 2004). The ITRS predicts that air gaps (AG) will be the most preferred medium between interconnects in future technologies beyond 16 nm. However, fabrication of Cu interconnects with air gaps is process intensive, consumes large chip area and is a costly affair (Shamiryan et al. 2004). In view of the advancements in CNT interconnect technology for beyond 16 nm technology nodes (Ceyhan and Naeemi 2013), CNT/AG interconnects are proposed as the ideal candidates for future interconnect technologies which can outperform Cu/low-k interconnects. However, one of the main concerns in air-gap technologies is that the heat dissipated by the wires due to joule heating will get trapped in the air gaps which, in turn, may affect the performance of interconnects. This problem will be more severe for local interconnects which are smaller in dimensions, and hence may have smaller thermal conductivities. So, the extent up to which CNT/AG local interconnects, as depicted in Fig. 5.1, can operate at elevated

P. U. Sathyakam and P. S. Mallick, *Design and Crosstalk Analysis in Carbon Nanotube Interconnects*, https://doi.org/10.1007/978-981-15-8888-4_6

temperatures and the impact on their RC delay performance based on the temperature coefficient of resistance (TCR) of carbon nanotubes have been studied. The optimal interconnect diameter and lengths for local interconnects are determined for CNT/AG interconnects. The bias- and temperature-dependent resistance of a CNT is given by

$$R(V,T) = R_C + \frac{h}{4e^2}\frac{L + \lambda_{eff}(V,T)}{\lambda_{eff}(V,t)} \tag{6.1}$$

where R_C is the contact resistance, $h/4e^2$ is the intrinsic quantum resistance equal to 6.45 kΩ, L is length of interconnect in μm, λ_{eff} gives information about the acoustic (AC) and optical (OP) phonon mean free path (MFP) at room temperature and explained in detail by Pop et al. (2006). This equation is valid for lengths greater than effective mean free path of the interconnects. For ballistic lengths, the contact resistance dominates over the overall resistance of CNTs. Moreover, it gives information of the resistance as a function of bias-dependent MFP and the phonon MFP at room temperature. However, it must be noted that the resistance of a CNT interconnect is dependent on the ambient temperature as well, which is not included in Eq. 6.1. First, the ambient temperature-dependent resistance R(T) is extracted from CNT interconnects that affect the R(T)C delay. The temperature coefficient of resistance (TCR), which is basically the change of resistance per one degree celsius of temperature, is included in the calculations of R(T). Next, analysis of the R(T)C delay performance of CNT interconnects is done. This is essential for CNT/air-gap interconnects as the TCR-based thermal analysis can reveal the stability of CNT wires at elevated ambient temperatures. The temperature coefficient of resistance (TCR) α is the factor that determines the change in resistance of the conductor with varying temperature. For SWCNTs of lengths comparable to the low bias acoustic (AC) and optical (OP) phonon mean free path (MFP) at room temperature (L_0) and for lengths far greater than the phonon MFP, α is modelled as (Naeemi and Meindl 2007c)

$$\alpha = \begin{cases} \frac{(L/10^3 D/T_0)}{1+(L/10^3 D)(T/T_0-2)} & \text{for } L_0 \approx 10^3 D \\ \frac{1}{T-200K} & \text{for} L \gg L_0 \end{cases} \tag{6.2}$$

where L is the interconnect length in μm, D is CNT diameter in nm, T_0 is room temperature in kelvin (=300 K) and T is the temperature at which α is measured. So, the resistance of the interconnect can be written as

$$R(T) = R_0(T)(1 + \alpha(T - T_0)) \tag{6.3}$$

where R_0(T) is the resistance of CNT at room temperature as described in Eq. 6.1 and R(T) is the resistance measured at temperature T. Figure 6.1 shows increase in resistance $R(T)/R_{300K}$ with temperature, for D = 2 nm and L = 1 μm. Resistance, R(T), becomes 1.2 times the room temperature resistance at 400K. Figure 6.2 depicts the change in R(T)/R_{300K} at various diameters when L = 2μm as computed from Eq. 6.2. It is seen that the diameter is inversely proportional to the rise in resistance.

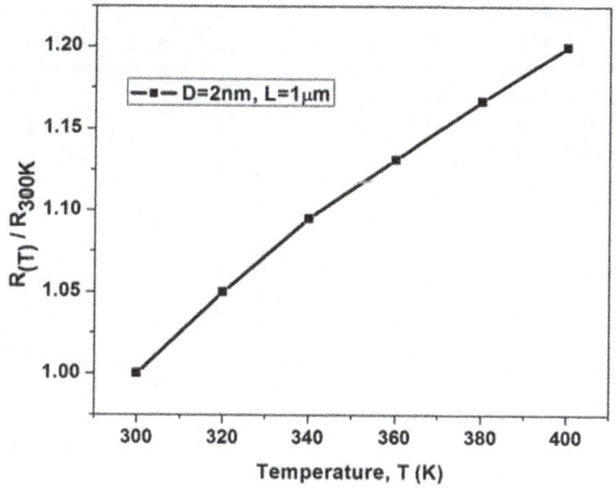

Fig. 6.1 Change of normalized resistance (R(T)/R(300K)) with increasing T for D = 2 nm, and L = 1 μm (Sathyakam and Mallick 2017)

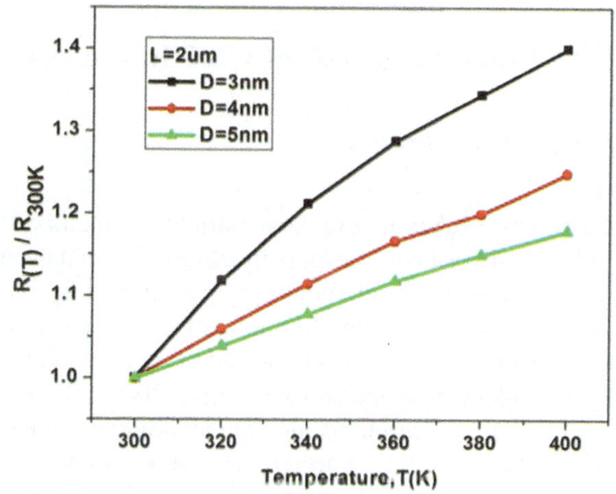

Fig. 6.2 Change of normalized resistance (R(T)/R(300K)) with increasing T for various diameters at L = 2 μm (Sathyakam and Mallick 2017)

Further, for larger lengths greater than the phonon MFP, R(T) reaches a maximum value of 10.7 kΩ for CNTs of length 6 μm and 6nm diameter as shown in Fig. 6.3. Local interconnects can have lengths up to 10 μm at low bias ($V < 0.3$ V) if one needs to restrict D within a certain limit. We further find that CNTs with diameters of 4 to 6 nm show better RC delays as compared to 3 nm CNTs at high temperatures. So one can increase the CNT diameter that is permissible within the technology node requirements. Further, greater number of conducting channels can offset the increase in *T* by decreasing the resistance due to negative TCR (Naeemi and Meindl 2007c). The CNT breakdown voltage increases with the input voltage as well as the ambient *T* for various lengths of the CNTs (Chen et al. 2009).

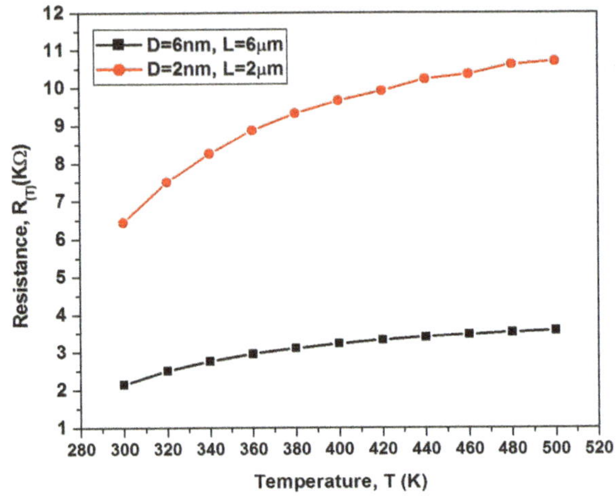

Fig. 6.3 Change in resistance at various temperatures for longer CNT interconnects ($L \gg L_0 = 1000 \times D$) (Sathyakam and Mallick 2017)

6.2 Fractal Analysis of CNT Interconnects

6.2.1 Introduction

Carbon nanotubes are excellent thermal conductors (Pop et al. 2006). They are found to be ideal for heat transport in nanoscale devices and interconnects. The thermal behaviour of CNT interconnects was studied by many researchers (Mohsin et al. 2013, 2014; Naeemi and Meindl 2007c). Recently, temperature-dependent analysis of CNT interconnects was done and the conductivity of a metallic CNT was found to be dependent on temperature (Pop et al. 2006). However, all these works have considered CNTs as continuous medium and did not include the discontinuous behaviour of heat flow in CNTs. Further, the porous structure of CNTs is ignored and the heat conduction was mainly modelled by the one-dimensional Joule's heat equation. As CNTs are porous materials, analysis of heat transfer in porous nanoscale media is very complex (Majumder et al. 2005).

Careful understanding of various physical phenomena in nanostructures is very important for real-world applications. For that, nanomaterials must be treated as discrete media. For example, water flow in carbon nanotubes was found to be five to six orders of magnitude faster than conventional flow in bulk media (Majumder et al. 2005). The main reason behind this interesting fact is attributed to the frictionless and hydrophobic walls of single-walled carbon nanotubes. This can be useful in applications like drug delivery.

This section analyses the heat conduction and flow in CNTs which are nanoscale discrete porous materials. For that, CNTs are considered as fractal media and fractional differential equations are applied that can explain the heat flow in CNTs.

Fig. 6.4 Sierpinski triangle

Fig. 6.5 Koch curve

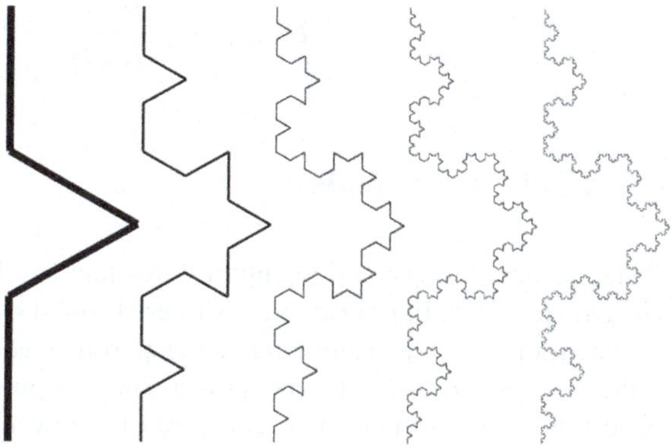

6.2.2 Introduction to Fractals

A fractal is a structure of self-similarity which repeats itself in a never-ending pattern. In some cases, the repetition can be finite. A classic example of fractals is the fractal triangle called Sierpinski triangle (Fig. 6.4).

In this triangle, the number of coloured triangles increases by a factor of 3 in each step of 1, 3, 9, 27, 81, 243, 729 and so on. Another example is the Koch curve as shown in Fig. 6.5 where the generator shape is repeatedly replaced by a similar copy of the shape, thereby increasing the overall length in each step. However, fractals can be complex. To understand and interpret complex fractals, fractional calculus is used that describes the fractional order of the fractal structure. By doing so, various physical phenomena like fluid flow, heat transfer, etc. can be modelled accurately.

6.2.3 Fractal Dimension

The fractal dimension (D) is the dimension of the fractal structure that gives the fractal order of the differential equation describing the fractal. Basically, it is the ratio of the number of self-similar pieces to the magnification factor of each piece. It is given as

$$D = \frac{log(\text{number of self-similar pieces})}{log(\text{Magnification factor})} \tag{6.4}$$

For example, a square can be divided into four (N^2) self-similar pieces with a magnification factor of 2 (N). So, D = logN^2/logN = 2. Similarly, a cube can be divided into eight (N^3) self-similar pieces where each piece has a magnification factor of 2 (N) which yields a D of log(N^3)/logN = 3. So, finally, the fractal dimension of a Sierpinski triangle is shown in Fig. 6.4 which has three self-similar pieces with a magnification factor of 2 and is given as

$$D_{Sirpinski} = \frac{log3}{log2} = 1.58 \tag{6.5}$$

6.3 CNTs as Fractals

CNTs are one-dimensional quantum wires that can be considered as fractal media (Hu and He 2016). Furthermore, CNTs are formed by sheets of hexagonally bonded carbon atoms. So, they can be treated as porous media with definite pore size. Heat conduction in porous fractal media is a complex process. CNTs showed very high-speed fluid flow through them compared to conventional flow speeds (Majumder et al. 2005). It can be five to six times higher in fractal media like CNTs than in bulk medium. Similarly, heat conduction in fractal media is analysed from the local temperature of the media which is dependent on the fractal dimensions as found in earlier examples of heat flow analysis (Liu, He and Li 2014).

In the case of CNTs, the heat flow must be first derived for the unit cell as shown in Fig. 6.6 and can be extended to the whole structure. Then the fractional dimensions must be applied to the partial derivatives to find the exact heat conduction in the CNT. The fractal dimension of an armchair (n,n) SWCNT along its length (x-axis) is modelled as

$$D_s = \frac{log(m.l_0 \times 2n)}{log(m)} \tag{6.6}$$

where m is the number of unit cells in a single row for a length of 10 Å of SWCNT, l_0 is the length of the CNT $\geq$ 10Å and n is the chirality index. From the above equation, it must be noted that the fractal dimension of a CNT is dependent on its diameter as well as its length. So, for larger CNTs, the fractal dimension calculation becomes more complex as the number of self-similar cells increases. This was shown by Liu, He and Li (2014) where the heat conduction in nanoscale was modelled using a fractional differential equation, given as

$$K_x \frac{\partial^{2\alpha_1}}{\partial x^{2\alpha_1}} + K_y \frac{\partial^{2\alpha_2}}{\partial y^{2\alpha_2}} + K_z \frac{\partial^{2\alpha_3}}{\partial z^{2\alpha_3}} - \rho c \frac{\partial T}{\partial t} = 0 \tag{6.7}$$

where T is temperature, α_1, α_2 and α_3 are the fractional dimensions in x-, y- and z-axes, respectively. Kx, Ky and Kz are the thermal conductivity coefficients in x-, y- and z-axes, respectively. Since CNTs are considered as long quantum wires,

Fig. 6.6 Unit cell of a CNT is within the dotted lines; a_1 and a_2 are basis vectors

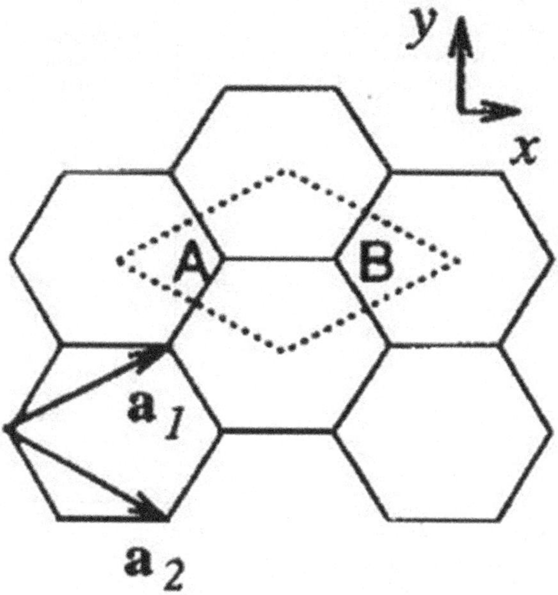

one can safely neglect the third dimension in the above equations. While one can consider the heat transfer as a time-dependent process, there are certain cases where the heat flow is time independent which can be the case for a heterogeneous or non-isotropic medium. A more detailed analysis of fractal derivatives and various forms of fractional calculus can be found in (Gabano and Poinot 2009; He 2014; Sierociuk et al. 2013). Furthermore, a fractal derivative, which captures the fluid flow in a porous medium that has a pore length of L0, is given as

$$\frac{Du}{Dx^{\alpha}} = \lim_{\Delta x \to L_0} \frac{u(A) - u(B)}{k(x_A - x_B)^{\alpha}} \tag{6.8}$$

where α is the fractal dimension, k is a constant, and A and B are any two points on the sample (Hu and He 2016).

6.4 Results and Discussions

6.4.1 *Temperature-Dependent Analysis of CNT/AG Interconnects*

After calculating the temperature-dependent resistance of CNTs, the corresponding $R_{(T)}C$ delay is found out for different lengths and diameters of CNT/AG interconnects. Figures 6.7 and 6.8 show the $R_{(T)}C$ delay calculated.

Fig. 6.7 $R_T C$ delay of CNT/AG interconnects at different temperatures for L = 1 μm and varying CNT diameters (Sathyakam and Mallick 2017)

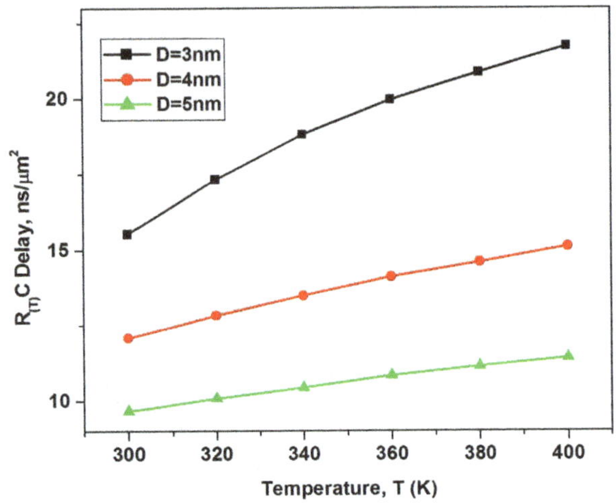

Fig. 6.8 $R_T C$ delay of CNT/AG interconnects at different temperatures for varying CNT lengths and diameters (Sathyakam and Mallick 2017)

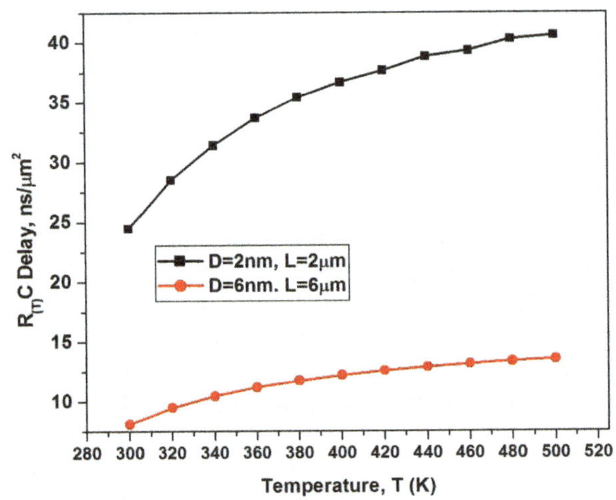

By considering the temperature-dependent resistance R(T), the $R_{(T)}C$ delay scales up with T as seen in Fig. 6.7. The slope is far steeper for small diameter CNTs as expected. This means that larger diameter CNTs should be considered at smaller technology nodes as well to overcome the thermal resistance problem. Further, for longer wires, the calculated delay for temperatures as high as 500K for two cases of CNT diameter and length is shown in Fig. 6.8. The delay is greater for small diameter CNTs.

Table 6.1 Fractal dimensions of various armchair SWCNTs

Armchair SWCNT (n,n)	Fractal dimensions	
	L = 10 Å	L = 15 Å
(2,2)	1.666	1.5579
(3,3)	1.8617	1.7211
(4,4)	2.00	1.8369
(5,5)	2.1074	1.9266

6.4.2 *Fractal Dimensions of Armchair SWCNTs*

Fractal dimensions of armchair CNTs that are used as interconnects are calculated from Eq. 6.6. Table 6.1 shows the fractal dimensions calculated for typical armchair SWCNTs of lengths 10 and 15 Å.

Overall, the range of fractal dimension is $1 < D_s < 3$ for an SWCNT. The length dependence shows that as the length increases, the fractal dimension approaches to 1 for a fixed diameter. On the other hand, as the diameter increases, the fractal dimension approaches to 3 for a fixed length. So, for long CNTs greater than 20 Å, the fractal dimension is near to unity, and hence the heat transport can be considered along the length of the CNT only.

6.4.3 *Summary*

Temperature coefficient of resistance (TCR) calculations of CNT interconnects shows that the resistance of CNT bundles increased with rise in ambient temperature. Our initial analysis of RC delay shows that it remains almost constant for all the technology nodes at 50 ns. Further, the R(T)C delay was found from SmartSPICE simulations which decreases with increase in CNT diameter and length and increases with temperature. The fractal dimensions of a SWCNT are modelled and its values are calculated for two different lengths of SWCNTs. It was found that as the length of the SWCNT increases, the fractal dimension approaches unity, and hence the heat flow is more prominent along the length of the SWCNT than its diameter.

Chapter 7
High-Speed CNT Interconnects

A case study of design and modelling of high-speed CNT interconnects is presented in this chapter. CNTFET-based driver/load circuits are designed and the channel length of the CNTFETs used is adjusted such that maximum current flows through the circuit and drives the interconnect. The analysis is carried at subthreshold conditions and the best cases of driver/load circuits are found out among six combinations of inverters and transmission gates. Lastly, temperature-dependent RLC models are discussed and the performance is found at different interconnect operating temperatures.

7.1 Introduction

Subthreshold voltage-operated circuits are the future for ultra-low-power applications. These circuits are inherently slow due to the very small subthreshold currents. The performance of these circuits is characterized by the speed at which they operate and the amount of power they dissipate. For scaled-down ICs, reducing the supply voltage (V_{dd}) to subthreshold levels is inevitable to achieve low-power operation. But, smaller V_{dd} makes the circuits slower as the circuit RC delay becomes dominant.

Recent advancements in CNT interconnect design for subthreshold circuits show that single-walled CNTs are suitable as local interconnects as they have less delay due to small wire capacitance Jamal and Naeemi (2011). But they are driven by big transistors with a W/L ratio of 5 for nFETs and 10 for pFETs which makes the driver resistance dominant. Again, for global interconnects the driver width was changed to make the driver resistance higher than the interconnect resistance Pable and Hasan (2012). Basically, this technique is used to increase the subthreshold output current of the drivers. In the case of CNT bundle interconnects, the resistance was decreased by increasing the number of metallic CNTs in the bundle at the cost of increased capacitance Pable and Hasan (2012).

P. U. Sathyakam and P. S. Mallick, *Design and Crosstalk Analysis in Carbon Nanotube Interconnects*, https://doi.org/10.1007/978-981-15-8888-4_7

However, these conventional methods cannot enhance the performance of subthreshold circuits at scaled technologies. The idea of up-scaling transistors for higher resistance does not go well with *scaling-down* approach (Pable et al. 2013; Pable and Hasan 2012). Further, the overall path delay of the circuit remains high as the driver is optimized for higher output current and not for lesser delay. This leads to the slower performance of circuits at subthreshold voltages than at superthreshold voltages. Recent findings from fabrication of graphene nanoribbon interconnects for subthreshold FPGAs show that they operate at about 150 KHz only Lee et al. (2013). For long global interconnects, buffer insertion is not feasible when they conduct subthreshold current. Pable et al. (2013) suggested dynamic threshold MOSFETs (DTMOS) as drivers to avoid use of repeaters. But DTMOS consumes high dynamic power and is not suitable for low-power operation. High-speed operation and low-power consumption in scaled-down circuits at subthreshold region can be achieved by using nanoscale transistors like FinFETs Sivasankaran and Mallick (2014), CNT-FETs Jafarzadehpour and Keshavarzian (2016) or GNRFETs connected by CNT interconnects. In this chapter, two approaches are proposed to tackle the issues of small subthreshold currents and high propagation delay in scaled integrated circuits. First, the channel length and width of the CNTFETs are modulated that are used in driver circuits. An optimal length and width of the FETs were found where the output subthreshold current is maximum so that the interconnects can conduct the input signals faithfully. This is the optimal width and length for least possible resistance and capacitance of the driver, respectively. Secondly, new driver and load circuits are designed using transmission gates (T-gates) along with inverters to drive CNT interconnects at subthreshold voltages. The pull-up/pull-down operations of the T-gates are utilized which gives a full swing of the output waveform as well as their lesser dynamic power consumption to advantage at subthreshold voltages. By doing so, the propagation delay is reduced thereby increasing the speed of the interconnects.

Next, temperature-dependent performance of subthreshold circuits is studied. Rai et al. (2016), Rai et al. (2017), Chiariello et al. (2010), Qi-Jun et al. (2015) have studied the performance of SWCNT bundle, mixed CNT bundles and multi-layer graphene nanoribbons (MLGNR) interconnects. In all their studies, temperature plays a major role in determining the propagation delay and power dissipated by the interconnects. The main reason behind this phenomenon is the temperature dependence of the mean free path (MFP) which in turn increases the wire resistance. So, both temperature-independent and temperature-dependent analyses of subthreshold interconnects are carried out to find out the impact of temperature on their performance.

7.2 CNTFET Current Model

CNTFET is regarded as the most advanced device which has similarities of MOSFET but has an advantage of working in ballistic regime (Bozorgmehr et al. 2017; Jafarzadehpour and Keshavarzian 2016). The threshold voltage in a CNTFET is dependent on the diameter of the CNTs used in the channel and is expressed as

$$V_{th} \approx \frac{E_g}{2e} = \frac{\alpha_0 V_\pi}{eD_{CNT}} \approx \frac{0.43}{D_{CNT}} \tag{7.1}$$

where e is the unit electron charge, E_g is the CNT bandgap, V_π is the carbon $\pi - \pi$ bond energy and D_{CNT} is the diameter of the CNT given by

$$D_{CNT} = \frac{\sqrt{3}a_0\sqrt{n_1^2 + n_1n_2 + n_2^2}}{\pi} = 0.0783\sqrt{n_1^2 + n_1n_2 + n_2^2} \tag{7.2}$$

where (n_1, n_2) is the chiral number and a_0 is the carbon–carbon bond length which is 0.142 nm (Bozorgmehr et al. 2017). For a CNT with chiral indices (n_1, n_2) as (0, 19), its diameter is 1.5 nm and the threshold voltage is 0.3 V. The ON current in a CNTFET is influenced by the channel length L_{ch}. It can be expressed as

$$I_{CNTFET} = \frac{Ng_{CNT}(V_{dd} - V_{th})}{1 + g_{CNT}L_{ch}\rho_S} \tag{7.3}$$

where N is number of CNTs in the channel, g_{CNT} is the transconductance per CNT and ρ_S is the resistivity of the CNT channel. This is the subthreshold current of the CNTFET that is used in the design of driver and load circuits in this work.

7.3 Transistor Sizing for Subthreshold Operation

The overall RC delay of the circuit depends on the driver resistance (R_{driv}) and capacitance (C_{driv}). By changing the size of the length and width of the transistor, one can change the RC delay as well as the output current (subthreshold current) of the driver (Calhoun et al. 2005). The threshold voltage in a circuit is reduced when the width of the transistors is reduced. This is called inverse narrow width effect (INWE) (Akers 1986). Ideally, minimum-sized drivers are necessary for high-speed and low-power circuits. However, it must be noted that the current is not minimum when the width is minimum (Muker and Shams 2011).

As per Eq. 7.3, I_{CNTFET} is dependent on transistor channel length. However, traditionally, sizing of the transistor was mainly done by decreasing the width and not the length, for subthreshold operation (Pable and Hasan 2012; Jamal and Naeemi 2011; Pable et al. 2013). While scaling of the gate width reduces capacitance (given as $C_{CNTFET} = N.C_{CNT} + C_{gate}$ (Jafarzadehpour and Keshavarzian 2016)), scaling of channel length reduces resistance. The expression for the width of the gate of the CNTFET is given by

$$W_{gate} = Min(W_{min}, N_g.pitch) \tag{7.4}$$

where W_{min} is the minimum width of the gate and N_g is the number of nanotubes under the gate. So, by varying both the length and width of the transistor, it is to reduce the driver RC delay, and hence increase the speed of the circuit. CNTFETs are used for constructing the driver and load circuits for subthreshold operation of

CNT interconnects. Recent work on CNTFET device fabrication shows that they are ideal for subthreshold operation (Franklin and Chen 2010; Franklin et al. 2012). Scaling of channel length and contact length of the device to as low as 15 nm and 20 nm, respectively, was found to be feasible without any short channel effects (SCE) and drain-induced barrier lowering (DIBL) effects. Further, the drain current Id for 15 nm, 300 nm and 3 m channel lengths is around 5 A, 3 A and 1.5 A, respectively, for an input voltage of 0.25 V (Sivasankaran and Mallick 2014). The relation between the channel length and the total resistance, Rtot, of the device can be given as

$$R_{tot} = \frac{h}{2e^2 M} \frac{L-ch}{L_{mfp}} + 2R_C \tag{7.5}$$

where $h/2e^2$ is the quantum resistance per channel of a CNT, L_{ch} is the channel length, L_{mfp} is the mean free path for phonon scattering, M is the number of modes per channel, $M = 2$ for a CNT and R_C is the contact resistance. This is because the contact resistance is dependent on the length of the nanotube–metal contact and the diameter of the CNT in a CNTFET is given by

$$R_C = \frac{\rho_C}{L_C.D_{CNT}} \tag{7.6}$$

For subthreshold operations, it is necessary that the contact resistance must be as low as possible to maintain a good subthreshold swing. So, the contact length, LC, must be increased to get a lower RC. For around 50 nm contact length, the contact resistance will be around 10 k as modelled. The shortest channel length of 15 nm showed highest room temperature conductance of $0.7G_0$ (where G_0 is the conductance of a CNT given as $2e^2/h$) and transconductance of 40 µS. The channel resistance was measured from 9 to 11 kΩ which is closer to the nanotube quantum resistance of 6.45 kΩ. So, these effects of nanoscale CNTFETs are taken as advantage and they are used in driver circuit design. The inverter is made by VS-CNTEFT model library provided by Stanford University (https://nano.stanford.edu/stanford-cnfet2-model). By varying the channel lengths of n-CNTFET and p-CNTFET, it is possible to modulate its channel resistance and identify the point where maximum subthreshold current flows through the driver circuit. Figure 7.1 shows the subthreshold current for different transistor channel lengths of the CNTFET-based inverter circuit. It can be seen that the current is maximum at 15 nm channel length which is in good agreement with Sivasankaran and Mallick (2014). Even though the current reduces in the order of 0.1 A for higher lengths, it is always better to have maximum possible current in the circuit so that there is least propagation delay. Further, the devices have a very small inverse subthreshold slope of only 90 mV/decade which means that the devices can switch faster.

SPICE analysis of a CNT bundle interconnect driven by a CNTFET-based inverter was performed. The equivalent circuit model of the CNT interconnects considered here is described in Sect. 7.5. Figure 7.2a shows the driver-interconnect-load model of the simulation setup. A 0.3 V pulsed voltage is applied to the inverter. The output

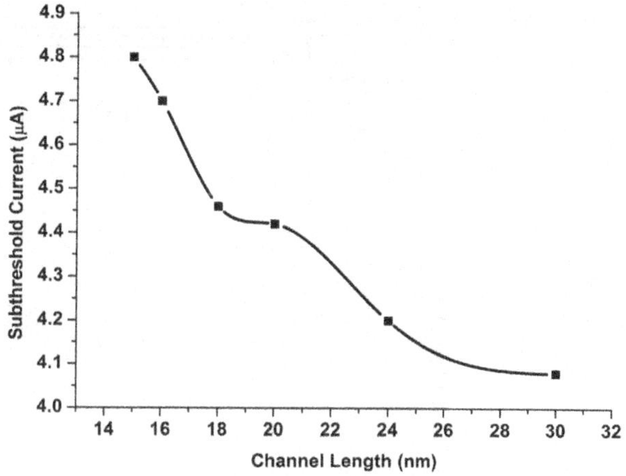

Fig. 7.1 Subthreshold current for different channel lengths of the CNTFETs in an inverter (Sathyakam et al. 2019)

voltage at the load side of the interconnect is plotted in Fig. 7.2b. It can be seen that the output waveform switches completely from V_{dd} to 0 V which is a major step towards high-speed operation of subthreshold currents. In order to capture the rising and falling trend of the output, the input waveforms are set with very short rise time or fall time which is 0.001 times smaller than time period. However, the delay of the circuit is still higher at 70 ns for high-speed operation of the circuits. The circuit worked at around 500 MHz which is roughly half of the speed at which current circuits work. So, to overcome the speed and delay issues, it is proposed here to use new driver circuits that are made of transmission gates and inverters. Ultimately, the advantage of scaling of the channel length of the devices as well as usage of transmission gates in the driver circuits of the subthreshold interconnects is utilized.

7.4 Proposed Circuit Models

7.4.1 CNTFET-Based Transmission Gate Model

Transmission gates (T-gates) are used in many applications like fast analog switch, logic gates, analog multiplexer in ADCs, control logic in programmable FPGAs, etc. Particularly, their high-speed switching action which is a result of good pull-up/down capability is utilized here to speed up subthreshold circuits. Typically, a transmission gate is made up of two transistors, an nFET and a pFET, connected in parallel as depicted in Fig. 7.3a and its circuit symbol is shown in Fig. 7.3b. The gates are connected to complementary inputs while the source takes the input signal and drain gives the output. If the nFET is fed with logic 1 and pFET with logic 0, the T-gate is ON, thereby passing the input signal to the output. For reverse logic, the T-gate is OFF.

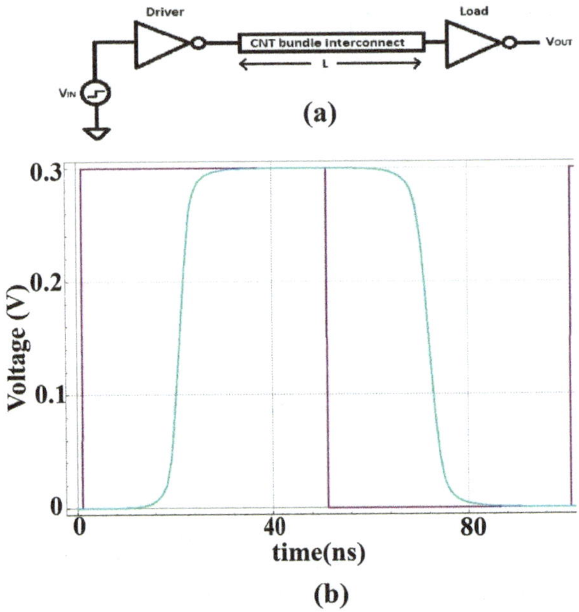

Fig. 7.2 **a** Proposed driver-interconnect-load model of CNTFET-based inverter driving a CNT interconnect (Case 1), **b** transient output waveform (Sathyakam et al. 2019)

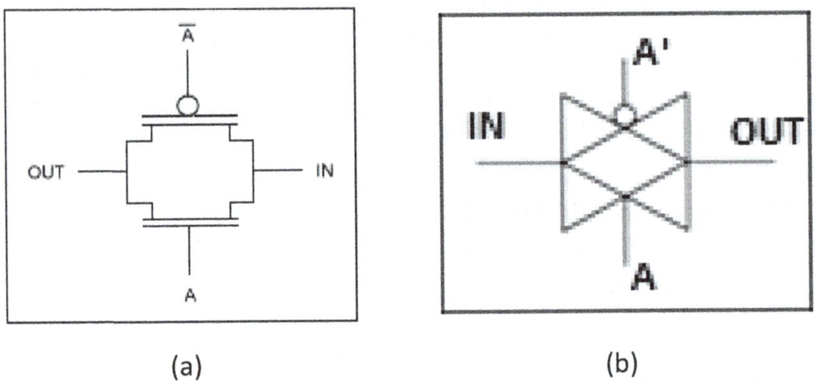

Fig. 7.3 **a** Transmission gate, **b** circuit symbol

The T-gates are constructed using CNTFET libraries. Various parameters of the model that are used in this work are enlisted in Table 7.1. The channel length of the pFET of the inverter, and pFET and nFET of the T-gate is taken as 15 nm where the current is maximum.

Table 7.1 Parameters from VS-CNTFET library

Parameter	Value
Subthreshold V_{DD}	0.3 V
Inverter nFET channel length (L_{ch})	10^{-24} nm
Chirality of CNTs	(19,0)
No. of CNTs in channel(N_g)	3
Diameter of each CNT	1.5 nm
Gate capacitance for $L_{ch} = 18$ nm	3.6 aF
Ion/Ioff for nFET	1.9×10^4
Ion/Ioff for pFET	2.8×10^4
Width of metal gate (W_{gate})	6.4 nm
CNT–metal contact resistance (R_C)	10 kΩ
Gate capacitance (C_{gate})	78 aF
Inverter pFET length	15 nm
T-gate nFET and pFET length	15 nm

7.4.2 *Different Configurations of Driver and Load Circuits*

A driver-interconnect-load (DIL) model is followed that describes the circuits made of inverters and T-gates as drivers and loads for CNT bundle interconnects. Throughout, the main motivation is to use the T-gate as a driver to pull up the signals to V_{dd} during ON state and to pull down the signals to 0 V during OFF state. Different cases of using T-gates along with the inverters in driver and load circuits are considered as shown in Fig. 7.4a–d. In case 1, a standard inverter is used as driver and load as shown in Fig. 7.2a. This circuit is taken as the reference for comparison purposes. The rest of the cases are shown in Fig. 7.4a–d. Case 2 is intended to use the T-gate to drive the subthreshold signals from the centre of the wire, i.e. it is used as a buffer at 500 m length. Next, in case 3, a T-gate is used just after the inverter so that both the inverter and T-gate can act as the driver when combined. In case 4, T-gate is placed before the load inverter. This is to study the effect of signals at the load when a T-gate is used at the load side. In case 5, two T-gates are used, one after the driver and another before the load, so that the signals at both driver side and load side can be analysed. The driver and load resistance is set by selecting the channel length of the n-CNTFET and the p-CNTFET. The channel length of both nFET and pFET is chosen to be 15 nm so that the channel resistance is minimum (Franklin and Chen 2010).

Normally, subthreshold signals are slow in transitioning from low to high and vice versa as the RC charging time of the interconnect is much larger than the signal transition time at high frequencies. This is the main limitation of subthreshold circuits when they operate at high frequencies. Nevertheless, it is a fact that device and interconnect scaling can be done simultaneously with voltage scaling to achieve better speed.

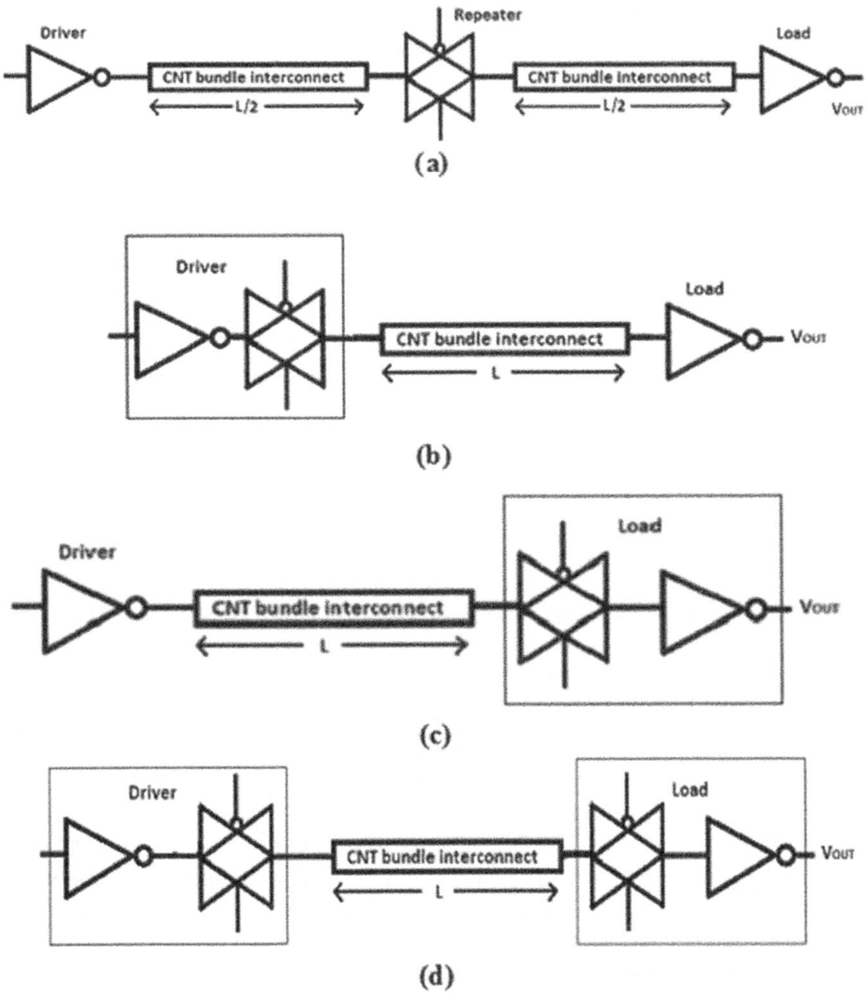

Fig. 7.4 Different DIL models considered in this work **a** Case 2: INV-wire-T-Gate-wire-INV model, **b** Case 3: INV-T-Gate-wire-INV model, **c** Case 4: INV-wire-T-Gate-INV model, **d** Case 5: INV-T-Gate-wire-T-Gate-INV model (Sathyakam et al. 2019)

7.5 CNT Bundle Interconnect Model

The transmission line equations of the voltage and current through an SWCNT bundle interconnect are given as

$$\frac{\partial V(z,t)}{\partial z} + L\frac{\partial I(z,t)}{\partial t} + RI(z,t) = 0 \tag{7.7}$$

$$\frac{\partial I(z,t)}{\partial z} + C\frac{\partial V(z,t)}{\partial t} = 0 \tag{7.8}$$

Considering an individual CNT in a SWCNT bundle, the number of conducting channels that contribute to its electrical conduction is given as

$$N_i = \begin{cases} aDT_i + b, & \text{if } D_i > d_T/T \\ 2, & \text{if } D_i < d_T/T \end{cases} \tag{7.9}$$

where T is temperature in kelvin, D_i is the diameter of the ith shell, $a = 3.87 \times 10^{-4}\,\text{nm}^{-1}\,\text{K}^{-1}$, b = 0.2 and $d_T = 1300\,\text{nmK}$. The interconnect resistance is dependent on its effective mean free path λ_{eff} (MFP) as explained by Rai et al. (2017). The factors that affect λ_{eff} are the optical phonon and acoustic phonon scattering mechanisms which have different scattering mean free paths at different ambient temperatures. The MFP due to acoustic phonons can be given as

$$\lambda_{AC} = \lambda_{AC.300}\left(\frac{300}{T}\right) \tag{7.10}$$

where $\lambda_{AC.300}$ is the acoustic phonon mean free path at 300 K which is roughly 1.6 m. The second contribution to λ_{eff} is the optical phonon scattering. So, both optical absorption ($\lambda_{OP.abs}$) and optical emission ($\lambda_{OP.ems}$) need to be considered which can be given as

$$\lambda_{OP.abs} = \lambda_{OP.300}\left(\frac{N_{OP}(300)+1}{N_{OP}(T)}\right) \tag{7.11}$$

$$\lambda_{OP.ems} = \frac{1}{\lambda^{fld}_{OP.ems}} + \frac{1}{\lambda^{abs}_{OP.ems}} \tag{7.12}$$

$$\lambda^{fld}_{OP.ems}(T) = \frac{\hbar\omega_{OP}}{eV_{dd}}l + \left(\frac{N_{OP}(300)+1}{N_{OP}(T)+1}\right)\lambda_{OP.300} \tag{7.13}$$

$$\lambda^{abs}_{OP.ems}(T) = \lambda_{OP.abs}(T) + \left(\frac{N_{OP}(300)+1}{N_{OP}(T)+1}\right)\lambda_{OP.300} \tag{7.14}$$

where $\lambda^{abs}_{OP.ems}$ is the optical emission MFP after absorption, $N_{OP}(T) = \frac{1}{[exp(\hbar\omega_{OP}/K_B T)-1]}$, $\lambda^{fld}_{OP.ems}(T)$ is the former optical emission MFP event, $\hbar_{OP}$ is the optical energy (0.18 eV) and $\lambda_{OP.300}$ is the spontaneous optical emission at 300 K which is roughly 15 nm. So, the effective MFP λ_{eff} can be given as

$$\frac{1}{\lambda_{eff}(T)} = \frac{1}{\lambda_{AC}} + \frac{1}{\lambda_{OP.abs}} + \frac{1}{\lambda_{OP.ems}} \tag{7.15}$$

Table 7.2 Interconnect parameters from equivalent single conductor model

Parameter	Value (per μm)
R_{ESC}	11.436 Ω
L_{kESC}	14.18 pH
C_{QESC}	225.6 fF
C_{eESC}	7.3157 pF

Considering these facts, the MFP and temperature-dependent resistance of a CNT bundle can be given as

$$R_b(T) = \frac{R_{SWCNT}}{n_B} = \begin{cases} \frac{R_C+R_Q}{n_B} & \text{for } l < \lambda_{eff} \\ \frac{R_C+R_S}{n_B} & \text{for } l > \lambda_{eff} \end{cases} \tag{7.16}$$

where n_B is the total number of CNTs in a bundle. The temperature-dependent expressions of the ESC inductance of a CNT bundle can be given as

$$L^b_{ESC}(T) = \frac{L_m + (L_k/N)}{n_B} \tag{7.17}$$

where L_m and L_k are the magnetic and kinetic inductances of an isolated CNT which are discussed in Chap. 2. Similarly, the effective capacitance of a CNT bundle placed on a ground plane is the series combination of the electrostatic and quantum capacitances of the CNT bundle and can be expressed as

$$C^b_{ESC}(T) = \frac{C^b_E \dot{C}^b_Q}{C^b_E + C^b_Q} \tag{7.18}$$

The electrostatic capacitance of the bundle is calculated by considering only the outer CNTs in the bundle as the inner CNTs are not capacitively coupled to the substrate due to the shielding effect of the outer CNTs. Table 7.2 gives the values of the ESC parameters computed from the above equations.

For 20 nm technology node, the width of the interconnect is 34 nm. Considering a square bundle of CNTs, the height is 34 nm. The total number of CNTs in the bundle is computed from the following equations.

7.6 Results and Discussions

7.6.1 *Transient Analysis*

First, transient analysis of the proposed DIL circuit models is performed using Silvaco SmartSPICE software. The required net-list codes are developed for each case of

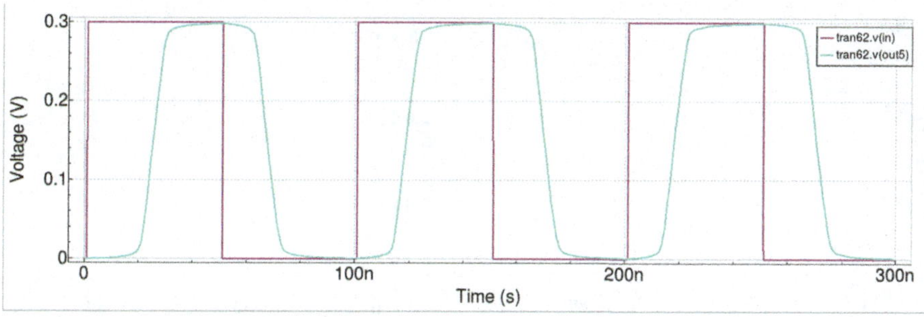

Fig. 7.5 Case 1: Inv-Wire-Inv

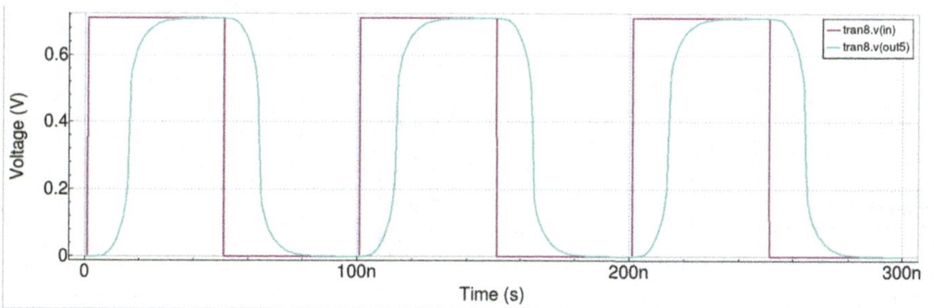

Fig. 7.6 Case 3: Inv-T-gate-Wire-Inv

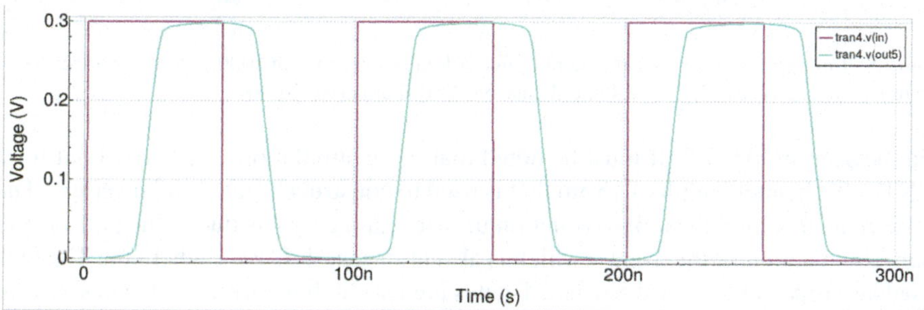

Fig. 7.7 Case 4: Inv-Wire-T-gate-Inv

driver/load configurations. V_{dd} is set to 0.3 V and rise time and fall time are set as 0.1ns. The time duration of ON state of the input signal is fixed at 20 ns and the time period is 40 ns. Figures 7.5, 7.6 and 7.7 show the waveforms for 500 μm interconnect length.

It can be seen from Fig. 7.5 that the delay is higher for an inverter-driven wire. But, as a T-gate is used at the driver side as in case 3, the delay is substantially reduced while maintaining the same swing, as seen from Fig. 7.6. But, when T-gate is used at the load side, the same speed enhancement as in the previous case is not observed as

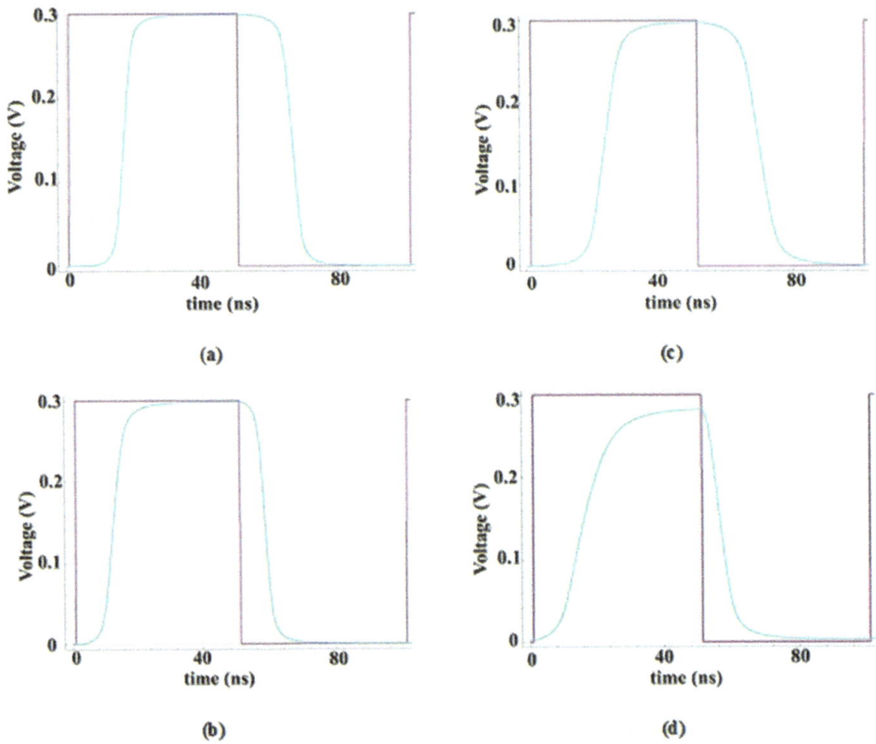

Fig. 7.8 Input (maroon colour) and output (blue colour) waveforms of 1000-μm-long wires, L_{ch} = 15 nm. **a** case 2, **b** case 3, **c** case 4 and **d** case 5 (Sathyakam et al. 2019)

can be seen in Fig. 7.7. It must be noted that these simulations are carried out for a CNTFET channel length of 15 nm that is used to construct T-gates and inverters. The main reason behind the delay reduction in case 3 and case 4 is due to the pull-up and pull-down action of the T-gate, which is absent in case 1. Case 2 where the T-gate is used as a repeater and case 5 where T-gates are used at both driver and load sides are not appropriate for 500 μm interconnects. So, their analysis is skipped here. Next, simulations are done for 1000 μm interconnects for cases 1 to 5 whose waveforms are shown in Fig. 7.8. The delay is highest for case 1 and least for case 5 followed by case 3. In case 5, the T-gates at both driver and load sides pulls up and down the waveform effectively. As a result, full swing is not achieved during ON state as the RC charging time is more than the time period of the input pulse. On the other hand, the delay is still lesser for case 3 due to the pull up of the input signal by the T-gate. Here, better ON state and OFF state are maintained. For case 2 and case 4, the delay is higher than case 3 and case 5. This is because, in case 2, the T-gate at the centre of the interconnect cannot effectively pull up the signal immediately after it enters the interconnect. For case 4, the T-gate at the load side cannot effectively pull up the signal at the input side. Hence, there is an increase in delay.

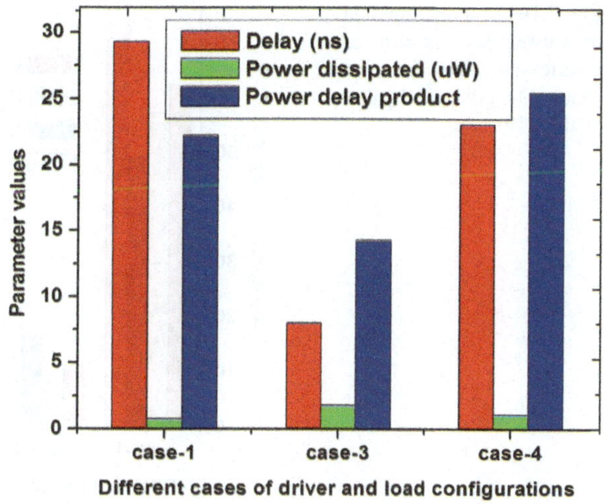

Fig. 7.9 Comparison of parameters for cases 1, 3 and 4 for interconnect length of 500 μm (Sathyakam et al. 2019)

7.6.2 Performance Analysis

The performance of subthreshold circuits is based on its operating speed, power dissipated by the circuit components and the reliability of the signals that pass through the circuit. The most important factor that limits the speed of a subthreshold voltage-driven circuit is its current density. At smaller V_{dd}, the current is smaller as compared to nominal V_{dd}-driven circuits. Hence, the resistance and the capacitance of the driver and load circuits are reduced by adjusting the transistor length. For an interconnect length of 500 μm, the performance of cases 1, 3 and 4 is compared. From Fig. 7.9, it can be observed that the delay is least for case 3. Case 4 has lesser delay than case 1. So, the best option is using T-gate at the driver side rather than in load side at 500 μm. This can be done at the expense of slight increase in power dissipation. The inverter and T gate transistor length is 15nm.

For a 1000-μm-long CNT bundle interconnect, the propagation delay for different cases of driver and load circuitry was found out. Also, the transistor length of the inverter is varied from 10 to 24 nm. Figure 7.10 shows the propagation delay for different cases from 1 to 5. As expected, the delay is lesser for cases 2, 3 and 5 where T-gates are used. Hence, the very reason for using T-gates in subthreshold circuits is to reduce delay. Particularly, in cases 2 and 3 where T-gates are used as drivers, the delay is least for 15 nm channel length of the inverter (L_{inv}) where the current is maximum. Delay increases as the length increases in cases 2 and 3. The delay improvements for case 3 with respect to cases 1, 2, 4 and 5 are 83%, 1.5%, 82% and 2%, respectively. In case 5, the delay is almost constant for all lengths. This is due to the fact that the T-gates at the driver side and load side are shielding the changes in the resistance of the inverter transistors in the driver as well as the load. This is very important as the process variations in the inverter circuitry are taken care by the T-gates while maintaining lesser propagation delay.

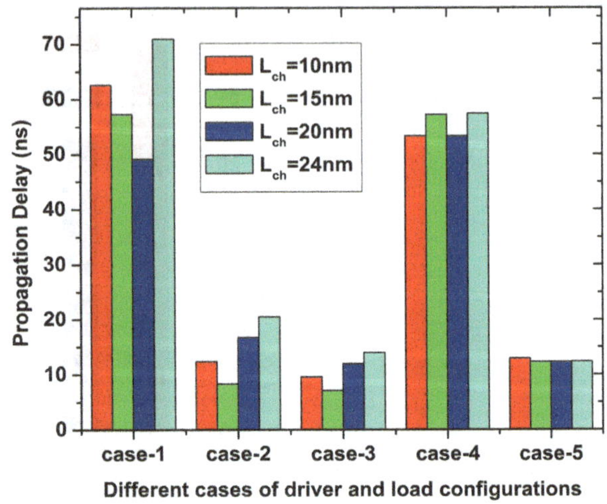

Fig. 7.10 Propagation delay for various cases at different inverter and T-gate transistor channel lengths. (Sathyakam et al. 2019)

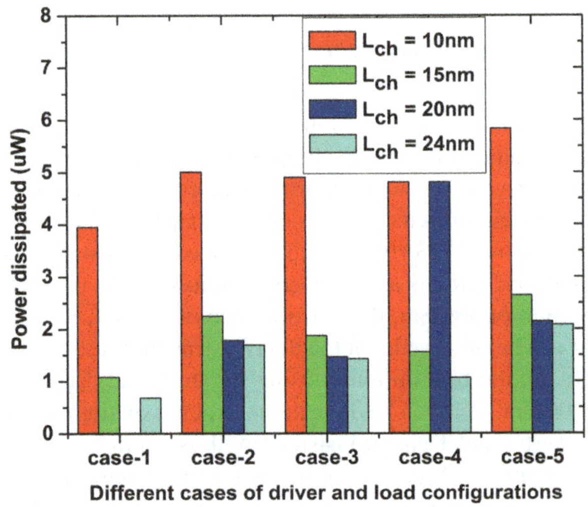

Fig. 7.11 Power dissipated for various cases at different inverter and T-gate transistor channel lengths. (Sathyakam et al. 2019)

The power dissipated by the circuits is shown in Fig. 7.11. Due to the high resistance of the transistors at Linv = 10 nm, the power dissipated by all the circuits is very high compared to other lengths. For case 1, the power dissipation analysis did not converge in a solution for Linv = 20 nm in the SPICE simulations. Further, the power dissipated in case 5, where T-gates are placed at driver side and load side, is highest. The power required by the load side T-gate to pull up and pull down the signal is very high as the output from the driver inverter is not passed through a T-gate immediately. However, it was found that the power dissipation for lengths other than 10 nm is 2%, 1%, 0.8% and 2.2% higher in cases 2, 3, 4 and 5, respectively, when compared to case 1. Obviously, this is due to the T-gates in the circuits. Next,

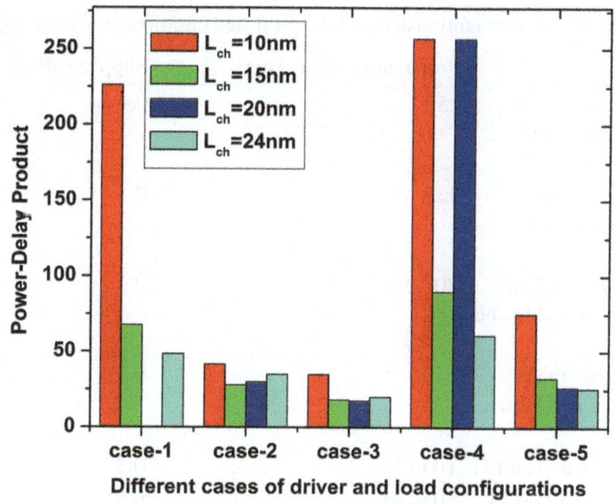

Fig. 7.12 Power-delay product for various cases and at different inverter and T-gate transistor channel lengths. (Sathyakam et al. 2019)

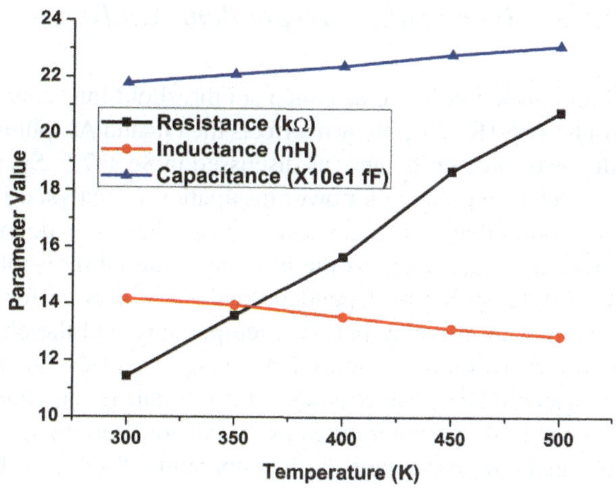

Fig. 7.13 Resistance, inductance and capacitance versus temperature of CNT interconnects (Sathyakam et al. 2019)

the power-delay product (PDP), which is a figure-of-merit for electric circuits, is calculated as depicted in Fig. 7.12. Overall, T-gate circuits have smaller PDP than inverter-based circuits. This shows the dominating nature of the propagation delay rather than the power dissipation on the performance of CNT interconnects. In each case, the PDP is higher for 10 nm length. Also, the PDP is highest in case 4 where T-gate is used at load side. The variation in PDP of case 4 is mainly due to the changes in the power dissipation at various lengths. The propagation delay of the proposed circuit models is compared with the model proposed by Pable and Hasan (2012) and shown in Table 7.3.

Table 7.3 Comparison of propagation delay of the proposed circuits along with suggested literature

Work	Propagation delay (ns)	Supply voltage (V)
Minimum wire spacing and DTMOS driver (Pable and Hasan 2012)	150	0.4
Total delay for min. wire spacing (Pable and Hasan 2012)	255	0.4
Tapered buffering technique (interconnect delay) (Pable and Hasan 2012)	20	0.4
Tapered buffering technique (path delay) (Pable and Hasan 2012)	140	0.4
This work (for cases 2, 3 and 5)	10 to 20	0.3
This work (for cases 1 and 4)	50 to 65	0.3

7.6.3 Temperature-Dependent Analysis

Ambient temperature at which subthreshold interconnects operate in ICs can be as high as 350 K. This, in turn, affects the OP and AC phonon mean free path, and hence the resistance of the wire as discussed in Sect. 7.5. So, their performance in terms of propagation delay and power dissipation is analysed in this section. However, as it was found that case 2, case 3 and case 5 have less delay, power dissipation and PDP, these cases are taken as the best cases for subthreshold operation of interconnects. Hence, temperature-dependent analysis was done for these cases only with $L_{ch} =$ 15 nm. The interplay between temperature and the channel length of the CNTFET can lead to more complex behaviour of carriers in the channel, and hence is not considered here. Nevertheless, it is felt that 15 nm channel length is a good estimate even at higher temperatures as the carrier lifetime and mobility will decrease as the channel length increases with temperature. Since a voltage of 0.3 V is applied, which is still higher than the low bias voltage of 0.16 V, both acoustic and optical phonon scattering mechanisms are dominant and are considered for the resistance calculations. From Fig. 7.13, the increase in the resistance of the interconnect due to the increased phonon scattering and reduced MFP can be noticed. The kinetic inductance is dependent on the number of conducting channels which is again dependent on temperature as given by Eq. (7.16). So, as the number of conducting channels increases with temperature, the inductance decreases as per Eq. (7.17). The magnetic inductance is negligible and is temperature independent. The quantum capacitance is again dependent on temperature and the number of conducting channels. So, it increases as the number of conducting channels increases due to temperature. After incorporating the temperature-dependent resistances, capacitance and inductance in the SPICE simulator, the propagation delay is found out which increases with increase in tem-

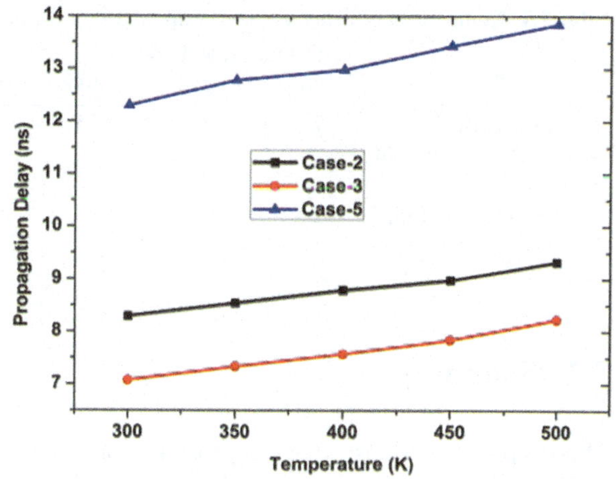

Fig. 7.14 Propagation delay of circuits in cases 2, 3 and 5 at various temperatures of CNT interconnects (Sathyakam et al. 2019)

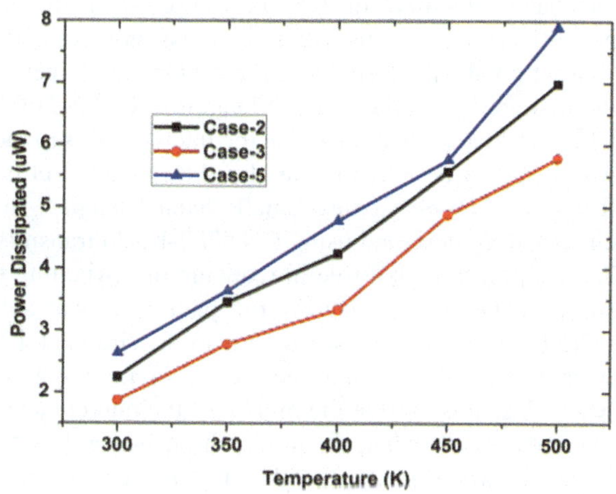

Fig. 7.15 Power dissipated by circuits in cases 2, 3 and 5 at various temperatures of CNT interconnects (Sathyakam et al. 2019)

perature as shown in Fig. 7.14. As a consequence, the power dissipated by the circuits increases with temperature as shown in Fig. 7.15. Among the three cases considered, the delay and power dissipation is least for case 3 compared to case 2 and case 5. As explained in the previous section, the main reason for the better performance of case 3 is due to the pull-up and pull-down action of the T-gate at the input side of the interconnect which drives the signals properly. This means that the high-speed performance of subthreshold CNT interconnects at elevated temperatures is possible with higher power dissipation.

Table 7.4 gives the comparison of temperature-dependent values of delay and power dissipated by the best case of this work and the work by Singh and Raj (2015b). In both the cases, the interconnect length is 1000 μm. Due to the difference in the applied voltage, the delay is higher for this work.

Table 7.4 Comparison of delay and power dissipated of the proposed circuits with literature

Work	Delay (ns) at T (K)			Power dissipation (μW) at T (K)		
	300	350	400	300	350	400
This work (case 2: best case) (subthreshold voltage = 0.3 V)	7	7.3	7.5	2	2.5	3
Singh and Raj (2015b) (nominal voltage = 0.8 V)	2.5	3	3.8	50	50	50

7.7 Summary

This chapter has discussed two approaches for enhancing the speed of CNT interconnects at subthreshold operating voltages. Both the approaches of (a) scaling down of the channel length of the CNTFETs used in the driver circuits and (b) designing of new driver/load circuits using transmission gates and inverters made of CNTFETs showed good enhancement in the subthreshold operation of CNT interconnects. In the first approach, the channel length of the CNTFETs was scaled down to 15 nm which yielded higher current in the interconnects to the tune of 5 μA. This approach was found to be attractive as the short channel effects and drain-induced barrier lowering is absent at such small channel lengths. In the second approach, driver circuits were designed using CNTFET-based transmission gates and inverters. The T-gates played a vital role in reducing the switching time, and hence enhances the speed of the circuits. The overall path delay was substantially reduced compared to traditional approaches. Analysis of the circuits for 500- and 1000-μm-long CNT interconnects showed that the propagation delay was enhanced by 83% for case 2 where T-gate is used at the middle of the interconnect, compared to inverter-driven interconnects. Further, the power dissipated is 1 to 5% higher for the cases where T-gates are used. Analysis of the PDP showed that the performance of CNT interconnects is more dependent on the propagation delay than the power dissipated by them. Finally, temperature-dependent analysis shows that the propagation delay and power dissipation increase drastically at high temperatures which show the limitations of the operation of subthreshold interconnects at high temperatures.

References

Abhyankar, N., Lee, M., Foley, M., Choi, E. S., Strouse, G., Kroto, H. W., & Dalal, N. S. (2016). Efficient synthesis and tailoring of magnetic and dielectric properties of pb-free perovskite-like abx3 metal-organic frameworks. *Physica Status Solidi (RRL)-Rapid Research Letters*, *10*(8), 600–605.

Akers, L. A. (1986). The inverse-narrow-width effect. *IEEE Electron Device Letters*, *7*(7), 419–421.

Alizadeh, A., & Sarvari, R. (2016). Temperature-dependent comparison between delay of cnt and copper interconnects. *IEEE Transactions on Very Large Scale Integration (VLSI) Systems*, *24*(2), 803–807.

Allendorf, M. D., Schwartzberg, A., Stavila, V., & Talin, A. A. (2011). A roadmap to implementing metal-organic frameworks in electronic devices: Challenges and critical directions. *Chemistry-A European Journal*, *17*(41), 11372–11388.

Amrin, S., & Deshpande, V. (2016). Mechanical and dielectric properties of carbon nanotubes/poly (vinyl alcohol) nanocomposites. In *AIP Conference Proceedings* (Vol. 1728, p. 020641). AIP Publishing.

Babaei, H., McGaughey, A. J., & Wilmer, C. E. (2017). Effect of pore size and shape on the thermal conductivity of metal-organic frameworks. *Chemical Science*, *8*(1), 583–589.

Bai, L., Yan, H., Yuan, L., & Liu, C. (2016). Synthesis of functionalized go for improving the dielectric properties of bismaleimide-triazine resin. *Journal of Polymer Research*, *23*(8), 169.

Baklanov, M. R., Adelmann, C., Zhao, L., & De Gendt, S. (2015). Advanced interconnects: Materials, processing, and reliability. *ECS Journal of Solid State Science and Technology*, *4*(1), Y1–Y4.

Baklanov, M. R., de Marneffe, J.-F., Shamiryan, D., Urbanowicz, A. M., Shi, H., Rakhimova, T. V., et al. (2013). Plasma processing of low-k dielectrics. *Journal of Applied Physics*, *113*(4), 4.

Banerjee, K., & Srivastava, N. (2006). Are carbon nanotubes the future of VLSI interconnections?. In *Proceedings of the 43rd Annual Design Automation Conference* (pp. 809–814). ACM.

Bellucci, S. (2005). Carbon nanotubes: Physics and applications. *Physica Status Solidi (c)*, *2*(1), 34–47.

Benedict, L. X., Louie, S. G., & Cohen, M. L. (1995). Static polarizabilities of single-wall carbon nanotubes. *Physical Review B*, *52*(11), 8541.

Bhattacharya, S., Das, D., & Rahaman, H. (2016). Reduced thickness interconnect model using GNR to avoid crosstalk effects. *Journal of Computational Electronics*, *15*(2), 367–380.

Biercuk, M. J., Ilani, S., Marcus, C. M., & McEuen, P. L. (2007). Electrical transport in single-wall carbon nanotubes. In *Carbon nanotubes* (pp. 455–493). Springer.

P. U. Sathyakam and P. S. Mallick, *Design and Crosstalk Analysis in Carbon Nanotube Interconnects*, https://doi.org/10.1007/978-981-15-8888-4

Blase, X., Rubio, A., Louie, S., & Cohen, M. (1994). Stability and band gap constancy of boron nitride nanotubes. *EPL (Europhysics Letters)*, *28*(5), 335.

Bockrath, M., Cobden, D. H., Lu, J., Rinzler, A. G., Smalley, R. E., Balents, L., et al. (1999). Luttinger-liquid behaviour in carbon nanotubes. *Nature*, *397*(6720), 598–601.

Bozorgmehr, A., Moayeri, M. H., Navi, K., & Bagherzadeh, N. (2017). Ultra-efficient fuzzy min/max circuits based on carbon nanotube fets. *IEEE Transactions on Fuzzy Systems*.

Burke, P. J. (2002). Luttinger liquid theory as a model of the gigahertz electrical properties of carbon nanotubes. *IEEE Transactions on Nanotechnology*, *99*(3), 129–144.

Calhoun, B. H., Wang, A., & Chandrakasan, A. (2005). Modeling and sizing for minimum energy operation in subthreshold circuits. *IEEE Journal of Solid-State Circuits*, *40*(9), 1778–1786.

Ceyhan, A., & Naeemi, A. (2013). Cu interconnect limitations and opportunities for SWNT interconnects at the end of the roadmap. *IEEE Transactions on Electron Devices*, *60*(1), 374–382.

Chae, H. G., & Kumar, S. (2006). Rigid-rod polymeric fibers. *Journal of Applied Polymer Science*, *100*(1), 791–802.

Chai, Y., Sun, M., Xiao, Z., Li, Y., Zhang, M., & Chan, P. C. (2011). Towards future VLSI interconnects using aligned carbon nanotubes. In *2011 IEEE/IFIP 19th International Conference on VLSI and System-on-Chip (VLSI-SoC)* (pp. 248–253). IEEE.

Chen, L.-Z., Pan, Q.-J., Cao, X.-X., & Wang, F.-M. (2016). Crystal structure, magnetism, and dielectric properties based on the axially chiral ligand 2, 2-dinitro-4, 4-biphenyldicarboxylic acid. *CrystEngComm*, *18*(11), 1944–1952.

Chen, W. C., Yin, W.-Y., Jia, L., & Liu, Q. H. (2009). Electrothermal characterization of single-walled carbon nanotube (SWCNT) interconnect arrays. *IEEE Transactions on Nanotechnology*, *8*(6), 718–728.

Chen, Y., Zou, J., Campbell, S. J., & Le Caer, G. (2004). Boron nitride nanotubes: Pronounced resistance to oxidation. *Applied Physics Letters*, *84*(13), 2430–2432.

Chiariello, A. G., Miano, G., & Maffucci, A. (2010). Size and temperature effects on the resistance of copper and carbon nanotubes nano-interconnects. In *2010 IEEE 19th Conference on Electrical Performance of Electronic Packaging and Systems (EPEPS)* (pp. 97–100). IEEE.

Chopra, N. G., Luyken, R., Cherrey, K., Crespi, V. H., et al. (1995). Boron nitride nanotubes. *Science*, *269*(5226), 966.

Ciofi, I., Contino, A., Roussel, P. J., Baert, R., Vega-Gonzalez, V.-H., Croes, K., et al. (2016). Impact of wire geometry on interconnect RC and circuit delay. *IEEE Transactions on Electron Devices*, *63*(6), 2488–2496.

Close, G. F., & Wong, H.-S. P. (2007). Fabrication and characterization of carbon nanotube interconnects. In *Electron Devices Meeting, 2007. IEDM 2007. IEEE International* (pp. 203–206). IEEE.

Close, G. F., Yasuda, S., Paul, B., Fujita, S., & Wong, H.-S. P. (2008). A 1 GHz integrated circuit with carbon nanotube interconnects and silicon transistors. *Nano Letters*, *8*(2), 706–709.

Cole, M. T., Cientanni, V., & Milne, W. I. (2016). Horizontal carbon nanotube alignment. *Nanoscale*, *8*(35), 15836–15844.

Croes, K., Wu, C., Kocaay, D., Li, Y., Roussel, P., Bömmels, J., et al. (2015). Current understanding of BEOL TDDB lifetime models. *ECS Journal of Solid State Science and Technology*, *4*(1), N3094–N3097.

D'amore, M., Sarto, M. S., & Tamburrano, A. (2010). Fast transient analysis of next-generation interconnects based on carbon nanotubes. *IEEE Transactions on Electromagnetic Compatibility*, *52*(2), 496–503.

Das, D., & Rahaman, H. (2011). Analysis of crosstalk in single-and multiwall carbon nanotube interconnects and its impact on gate oxide reliability. *IEEE Transactions on Nanotechnology*, *10*(6), 1362–1370.

Demczyk, B., Wang, Y., Cumings, J., Hetman, M., Han, W., Zettl, A., et al. (2002). Direct mechanical measurement of the tensile strength and elastic modulus of multiwalled carbon nanotubes. *Materials Science and Engineering: A*, *334*(1–2), 173–178.

Dhieb, A., Valkonen, A., Rzaigui, M., & Smirani, W. (2015). Synthesis, crystal structure, physico-chemical characterization and dielectric properties of a new hybrid material, 1-ethylpiperazine-1, 4-diium tetrachlorocadmate. *Journal of Molecular Structure*, *1102*, 50–56.

Fagan, J., Simpson, J. R., Landi, B., Richter, L., Mandelbaum, I., Bajpai, V., et al. (2007). Dielectric response of aligned semiconducting single-wall nanotubes. *Physical Review Letters*, *98*(14), 147402.

Farrell, R., Goshal, T., Cvelbar, U., Petkov, N., & Morris, M. A. (2011). Advances in ultra low dielectric constant ordered porous materials. *The Electrochemical Society Interface*, *20*(4), 39–46.

Fathi, D., & Forouzandeh, B. (2009a). A novel approach for stability analysis in carbon nanotube interconnects. *IEEE Electron Device Letters*, *30*(5), 475–477.

Fathi, D., & Forouzandeh, B. (2009b). Time domain analysis of carbon nanotube interconnects based on distributed RLC model. *Nano*, *4*(01), 13–21.

Fathi, D., Forouzandeh, B., Mohajerzadeh, S., & Sarvari, R. (2009). Accurate analysis of carbon nanotube interconnects using transmission line model. *Micro & Nano Letters*, *4*(2), 116–121.

Ferranti, F., Antonini, G., Dhaene, T., & Knockaert, L. (2014). Parametric macromodels for efficient design of carbon nanotube interconnects. *IEEE Transactions on Electromagnetic Compatibility*, *56*(6), 1674–1681.

Franklin, A. D., & Chen, Z. (2010). Length scaling of carbon nanotube transistors. *Nature Nanotechnology*, *5*(12), 858–862.

Franklin, A. D., Luisier, M., Han, S.-J., Tulevski, G., Breslin, C. M., Gignac, L., et al. (2012). Sub-10 nm carbon nanotube transistor. *Nano Letters*, *12*(2), 758–762.

Fuentes, G., Borowiak-Palen, E., Pichler, T., Liu, X., Graff, A., Behr, G., et al. (2003). Electronic structure of multiwall boron nitride nanotubes. *Physical Review B*, *67*(3), 035429.

Gabano, J.-D., & Poinot, T. (2009). Fractional modelling applied to heat conductivity and diffusivity estimation. *Physica Scripta*, *T136*, 014015.

Gambino, J. (2011). Process challenges for integration of copper interconnects with low-k dielectrics. *ECS Transactions*, *35*(4), 687–699.

Gao, W. (2015). The chemistry of graphene oxide. In *Graphene oxide* (pp. 61–95). Springer.

García-Fernández, A., Bermúdez-García, J., Castro-García, S., Artiaga, R., López-Beceiro, J., Señarís-Rodríguez, M., et al. (2016). Dielectric properties induced by the framework in the hybrid organic-inorganic compounds m (dca) 2pyz m= fe, co and zn. *Polyhedron*, *114*, 249–255.

Gebhardt, B., Syrgiannis, Z., Backes, C., Graupner, R., Hauke, F., & Hirsch, A. (2011). Carbon nanotube sidewall functionalization with carbonyl compounds modified birch conditions vs the organometallic reduction approach. *Journal of the American Chemical Society*, *133*(20), 7985–7995.

Gholipour, M., & Masoumi, N. (2011). Efficient model for delay estimation of MWCNT interconnects. In *2011 International Conference on Microelectronics (ICM)* (pp. 1–4). IEEE.

Giustiniani, A., Tucci, V., & Zamboni, W. (2011). Carbon nanotubes bundled interconnects: Design hints based on frequency-and time-domain crosstalk analyses. *IEEE Transactions on Electron Devices*, *58*(8), 2702–2711.

Guan, L., Suenaga, K., & Iijima, S. (2008). Smallest carbon nanotube assigned with atomic resolution accuracy. *Nano Letters*, *8*(2), 459–462.

Guo, H., Wang, M., Liu, J., Zhu, S., & Liu, C. (2016). Facile synthesis of nanoscale high porosity IR-MOFs for low-k dielectrics thin films. *Microporous and Mesoporous Materials*, *221*, 40–47.

Ha, H. W., Choudhury, A., Kamal, T., Kim, D.-H., & Park, S.-Y. (2012). Effect of chemical modification of graphene on mechanical, electrical, and thermal properties of polyimide/graphene nanocomposites. *ACS Applied Materials & Interfaces*, *4*(9), 4623–4630.

Haruehanroengra, S., & Wang, W. (2007). Analyzing conductance of mixed carbon-nanotube bundles for interconnect applications. *IEEE Electron Device Letters*, *28*(8), 756–759.

Hatton, B. D., Landskron, K., Hunks, W. J., Bennett, M. R., Shukaris, D., Perovic, D. D., et al. (2006). Materials chemistry for low-k materials. *Materials Today*, *9*(3), 22–31.

Hayashi, T., & Endo, M. (2011). Carbon nanotubes as structural material and their application in composites. *Composites Part B: Engineering*, *42*(8), 2151–2157.

He, J.-H. (2014). A tutorial review on fractal spacetime and fractional calculus. *International Journal of Theoretical Physics*, *53*(11), 3698–3718.

Hong, X., Wang, D., & Chung, D. (2016). Boron nitride nanotube mat as a low-k dielectric material with relative dielectric constant ranging from 1.0 to 1.1. *Journal of Electronic Materials*, *45*(1), 453–461.

Hu, Y., & He, J.-H. (2016). On fractal space-time and fractional calculus. *Thermal Science*, *20*(3), 773–777.

Huang, B., McGaughey, A., & Kaviany, M. (2007). Thermal conductivity of metal-organic framework 5 (mof-5): Part i. molecular dynamics simulations. *International Journal of Heat and Mass Transfer*, *50*(3–4), 393–404.

Huang, B., Ni, Z., Millward, A., McGaughey, A., Uher, C., Kaviany, M., et al. (2007). Thermal conductivity of a metal-organic framework (mof-5): Part ii. measurement. *International Journal of Heat and Mass Transfer*, *50*(3–4), 405–411.

Iijima, S., & Ichihashi, T. (1993). Single-shell carbon nanotubes of 1-nm diameter. *Nature*, **363**(6430), 603.

Iqbal, Q., Bernstein, P., Zhu, Y., Rahamim, J., Cebe, P., & Staii, C. (2015). Quantitative analysis of mechanical and electrostatic properties of poly (lactic) acid fibers and poly (lactic) acidcarbon nanotube composites using atomic force microscopy. *Nanotechnology*, *26*(10), 105702.

Jafarzadehpour, F., & Keshavarzian, P. (2016). Low-power consumption ternary full adder based on CNTFET. *IET Circuits, Devices & Systems*, *10*(5), 365–374.

Jamal, O., & Naeemi, A. (2011). Ultralow-power single-wall carbon nanotube interconnects for subthreshold circuits. *IEEE Transactions on Nanotechnology*, *10*(1), 99–101.

James, S. L. (2003). Metal-organic frameworks. *Chemical Society Reviews*, *32*(5), 276–288.

Jiang, J., Kang, J., Cao, W., Xie, X., Zhang, H., Chu, J. H., et al. (2017). Intercalation doped multilayer-graphene-nanoribbons for next-generation interconnects. *Nano Letters*, *17*(3), 1482–1488.

Joshi, G. M., & Deshmukh, K. (2016). Study of conjugated polymer/graphene oxide nanocomposites as flexible dielectric medium. *Journal of Materials Science: Materials in Electronics*, *27*(4), 3397–3409.

Kalay, S., Yilmaz, Z., Sen, O., Emanet, M., Kazanc, E., & Çulha, M. (2015). Synthesis of boron nitride nanotubes and their applications. *Beilstein Journal of Nanotechnology*, *6*, 84.

Kane, C., Mele, E., Lee, R., Fischer, J., Petit, P., Dai, H., et al. (1998). Temperature-dependent resistivity of single-wall carbon nanotubes. *EPL (Europhysics Letters)*, *41*(6), 683.

Kaushik, B. K., & Sarkar, S. (2008a). Crosstalk analysis for a CMOS-gate-driven coupled interconnects. *IEEE Transactions on Computer-Aided Design of Integrated Circuits and Systems*, *27*(6), 1150–1154.

Kaushik, B. K., & Sarkar, S. (2008b). Crosstalk analysis for a CMOS gate driven inductively and capacitively coupled interconnects. *Microelectronics Journal*, *39*(12), 1834–1842.

Kaushik, B. K., Sarkar, S., & Agarwal, R. P. (2007). Waveform analysis and delay prediction for a CMOS gate driving RLC interconnect load. *Integration, The VLSI Journal*, *40*(4), 394–405.

Koo, K.-H., Cho, H., Kapur, P., & Saraswat, K. C. (2007). Performance comparisons between carbon nanotubes, optical, and cu for future high-performance on-chip interconnect applications. *IEEE Transactions on Electron Devices*, *54*(12), 3206–3215.

Kozinsky, B., & Marzari, N. (2006). Static dielectric properties of carbon nanotubes from first principles. *Physical Review Letters*, *96*(16), 166801.

Kreupl, F., Graham, A. P., Duesberg, G., Steinhögl, W., Liebau, M., Unger, E., et al. (2002). Carbon nanotubes in interconnect applications. *Microelectronic Engineering*, *64*(1), 399–408.

Krupke, R., Hennrich, F., Kappes, M. M., & v. Lhneysen, H., (2004). Surface conductance induced dielectrophoresis of semiconducting single-walled carbon nanotubes. *Nano Letters*, *4*(8), 1395–1399.

Kumar, R., & Parashar, A. (2016). Atomistic modeling of bn nanofillers for mechanical and thermal properties: A review. *Nanoscale*, *8*(1), 22–49.

Kumar, V. R., Majumder, M. K., Alam, A., Kukkam, N. R., & Kaushik, B. K. (2015). Stability and delay analysis of multi-layered GNR and multi-walled CNT interconnects. *Journal of Computational Electronics*, *14*(2), 611–618.

Kumaresan, V., Wilson, C., Verdonck, P., Van Besien, E., Lazzarino, F., Truffert, V., et al. (2014). Simulation and measurement of the capacitance benefit of air gap interconnects for advanced technology nodes. *Microelectronic Engineering*, *120*, 90–94.

Lee, K.-J., Park, H., Kong, J., & Chandrakasan, A. P. (2013). Demonstration of a subthreshold FPGA using monolithically integrated graphene interconnects. *IEEE Transactions on Electron Devices*, *60*(1), 383–390.

Lee, S., & Lee, B.-J. (2012). Removal of residual oxide layer formed during chemical-mechanical-planarization process for lowering contact resistance. *Surface and Coatings Technology*, *206*(13), 3142–3145.

Li, H., & Banerjee, K. (2009). High-frequency analysis of carbon nanotube interconnects and implications for on-chip inductor design. *IEEE Transactions on Electron Devices*, *56*(10), 2202–2214.

Li, H., Liu, W., Cassell, A. M., Kreupl, F., & Banerjee, K. (2013a). Low-resistivity long-length horizontal carbon nanotube bundles for interconnect applications part i: Process development. *IEEE Transactions on Electron Devices*, *60*(9), 2862–2869.

Li, H., Liu, W., Cassell, A. M., Kreupl, F., & Banerjee, K. (2013b). Low-resistivity long-length horizontal carbon nanotube bundles for interconnect applications part ii: characterization. *IEEE Transactions on Electron Devices*, *60*(9), 2870–2876.

Li, H., Lu, W., Li, J., Bai, X., & Gu, C. (2005). Multichannel ballistic transport in multiwall carbon nanotubes. *Physical Review Letters*, *95*(8), 086601.

Li, H., Srivastava, N., Mao, J.-F., Yin, W.-Y., & Banerjee, K. (2011). Carbon nanotube vias: Does ballistic electron-phonon transport imply improved performance and reliability? *IEEE Transactions on Electron Devices*, *58*(8), 2689–2701.

Li, H., Xu, C., Srivastava, N., & Banerjee, K. (2009). Carbon nanomaterials for next-generation interconnects and passives: Physics, status, and prospects. *IEEE Transactions on Electron Devices*, *56*(9), 1799–1821.

Li, H., Yin, W.-Y., Banerjee, K., & Mao, J.-F. (2008). Circuit modeling and performance analysis of multi-walled carbon nanotube interconnects. *IEEE Transactions on Electron Devices*, *55*(6), 1328–1337.

Li, J., Ye, Q., Cassell, A., Ng, H. T., Stevens, R., Han, J., et al. (2003). Bottom-up approach for carbon nanotube interconnects. *Applied Physics Letters*, *82*(15), 2491–2493.

Li, W. J., Liu, J., Sun, Z.-H., Liu, T.-F., Lü, J., Gao, S.-Y., et al. (2016). Integration of metal-organic frameworks into an electrochemical dielectric thin film for electronic applications. *Nature Communications*, *7*, 11830.

Li, Y.-H., & Lue, J.-T. (2007). Dielectric constants of single-wall carbon nanotubes at various frequencies. *Journal of Nanoscience and Nanotechnology*, *7*(9), 3185–3188.

Liang, F., Wang, G., & Ding, W. (2011). Estimation of time delay and repeater insertion in multiwall carbon nanotube interconnects. *IEEE Transactions on Electron Devices*, *58*(8), 2712–2720.

Liang, F., Wang, G., & Lin, H. (2011). Modelling of self-heating effects in multi-wall carbon nanotube interconnects. *Micro & Nano Letters*, *6*(1), 52–54.

Liaw, D.-J., Wang, K.-L., Huang, Y.-C., Lee, K.-R., Lai, J.-Y., & Ha, C.-S. (2012). Advanced polyimide materials: Syntheses, physical properties and applications. *Progress in Polymer Science*, *37*(7), 907–974.

Lionti, K., Volksen, W., Magbitang, T., Darnon, M., & Dubois, G. (2015). Toward successful integration of porous low-k materials: Strategies addressing plasma damage. *ECS Journal of Solid State Science and Technology*, *4*(1), N3071–N3083.

Liu, B., Liu, J., Li, H.-B., Bhola, R., Jackson, E. A., Scott, L. T., et al. (2014). Nearly exclusive growth of small diameter semiconducting single-wall carbon nanotubes from organic chemistry synthetic end-cap molecules. *Nano Letters, 15*(1), 586–595.

Liu, H.-Y., He, J.-H., & Li, Z.-B. (2014). Fractional calculus for nanoscale flow and heat transfer. *International Journal of Numerical Methods for Heat & Fluid Flow, 24*(6), 1227–1250.

Lu, W., Wang, D., & Chen, L. (2007). Near-static dielectric polarization of individual carbon nanotubes. *Nano Letters, 7*(9), 2729–2733.

Ma, S., Wang, Y., Min, Z., & Zhong, L. (2013). Nano/mesoporous polymers based low-k dielectric materials: A review on methods and advances. *Advances in Polymer Technology, 32*(3).

Majumder, M., Chopra, N., Andrews, R., & Hinds, B. J. (2005). Nanoscale hydrodynamics: Enhanced flow in carbon nanotubes. *Nature, 438*(7064), 44.

Majumder, M. K., Das, P. K., & Kaushik, B. K. (2014). Delay and crosstalk reliability issues in mixed MWCNT bundle interconnects. *Microelectronics Reliability, 54*(11), 2570–2577.

Majumder, M. K., Kaushik, B. K., & Manhas, S. K. (2014). Analysis of delay and dynamic crosstalk in bundled carbon nanotube interconnects. *IEEE Transactions on Electromagnetic Compatibility, 56*(6), 1666–1673.

Majumder, M. K., Kumar, J., & Kaushik, B. K. (2015). Process-induced delay variation in SWCNT, MWCNT, and mixed CNT interconnects. *IETE Journal of Research, 61*(5), 533–540.

Majumder, M. K., Pandya, N. D., Kaushik, B., & Manhas, S. (2012a). Analysis of MWCNT and bundled SWCNT interconnects: Impact on crosstalk and area. *IEEE Electron Device Letters, 33*(8), 1180–1182.

Majumder, M., Pandya, N., Kaushik, B., & Manhas, S. (2012b). Dynamic crosstalk effect in mixed CNT bundle interconnects. *Electronics Letters, 48*(7), 384–385.

McEuen, P. L., Fuhrer, M. S., & Park, H. (2002). Single-walled carbon nanotube electronics. *IEEE Transactions on Nanotechnology, 99*(1), 78–85.

Mendiratta, S., Lee, C.-H., Usman, M., & Lu, K.-L. (2015). Metal-organic frameworks for electronics: Emerging second order nonlinear optical and dielectric materials. *Science and Technology of Advanced Materials, 16*(5), 054204.

Meo, M., & Rossi, M. (2006). Prediction of youngs modulus of single wall carbon nanotubes by molecular-mechanics based finite element modelling. *Composites Science and Technology, 66*(11), 1597–1605.

Moghaddam, S. A., Mohammadi, S., & Jabehdar-Maralani, P. (2013). Demystifying SWCNT-bundle-interconnects inductive behavior through novel modeling. *Journal of Computational Electronics, 12*(1), 1–13.

Mohsin, K. M., Srivastava, A., Sharma, A. K., & Mayberry, C. (2013). A thermal model for carbon nanotube interconnects. *Nanomaterials, 3*(2), 229–241.

Mohsin, K., Srivastava, A., Sharma, A. K., & Mayberry, C. (2014). Characterization of MWCNT VLSI interconnect with self-heating induced scatterings. In *IEEE Computer Society Annual Symposium on VLSI (ISVLSI)* (pp. 368–373). IEEE.

Muker, M., & Shams, M. (2011). Designing digital subthreshold CMOS circuits using parallel transistor stacks. *Electronics Letters, 47*(6), 372–374.

Myers, J. N., Zhang, X., Bielefeld, J. D., & Chen, Z. (2015). Plasma treatment effects on molecular structures at dense and porous low-k sicoh film surfaces and buried interfaces. *The Journal of Physical Chemistry C, 119*(39), 22514–22525.

Naeemi, A., & Meindl, J. D. (2005a). Impact of electron-phonon scattering on the performance of carbon nanotube interconnects for GSI. *IEEE Electron Device Letters, 26*(7), 476–478.

Naeemi, A., & Meindl, J. D. (2005b). Monolayer metallic nanotube interconnects: Promising candidates for short local interconnects. *IEEE Electron Device Letters, 26*(8), 544–546.

Naeemi, A., & Meindl, J. D. (2006). Compact physical models for multiwall carbon-nanotube interconnects. *IEEE Electron Device Letters, 27*(5), 338–340.

Naeemi, A., & Meindl, J. D. (2007a). Conductance modeling for graphene nanoribbon (GNR) interconnects. *IEEE Electron Device Letters, 28*(5), 428–431.

Naeemi, A., & Meindl, J. D. (2007b). Design and performance modeling for single-walled carbon nanotubes as local, semiglobal, and global interconnects in gigascale integrated systems. *IEEE Transactions on Electron Devices*, *54*(1), 26–37.

Naeemi, A., & Meindl, J. D. (2007c). Physical modeling of temperature coefficient of resistance for single-and multi-wall carbon nanotube interconnects. *IEEE Electron Device Letters*, *28*(2), 135–138.

Naeemi, A., Sarvari, R., & Meindl, J. D. (2005). Performance comparison between carbon nanotube and copper interconnects for gigascale integration (GSI). *IEEE Electron Device Letters*, *26*(2), 84–86.

Nasiri, S. H., Moravvej-Farshi, M. K., & Faez, R. (2010). Stability analysis in graphene nanoribbon interconnects. *IEEE Electron Device Letters*, *31*(12), 1458–1460.

Nieuwoudt, A., & Massoud, Y. (2006). Evaluating the impact of resistance in carbon nanotube bundles for VLSI interconnect using diameter-dependent modeling techniques. *IEEE Transactions on Electron Devices*, *53*(10), 2460–2466.

Nihei, M., Kondo, D., Kawabata, A., Sato, S., Shioya, H., Sakaue, M., Iwai, T., Ohfuti, M., & Awano, Y. (2005). Low-resistance multi-walled carbon nanotube vias with parallel channel conduction of inner shells [ic interconnect applications]. In *Interconnect Technology Conference, 2005. Proceedings of the IEEE 2005 International* (pp. 234–236). IEEE.

Ohba, N., Miwa, K., Nagasako, N., & Fukumoto, A. (2001). First-principles study on structural, dielectric, and dynamical properties for three bn polytypes. *Physical Review B*, *63*(11), 115207.

Ohya, H., Kudryavsev, V., & Semenova, S. I. (1997). In *Polyimide membranes: Applications, fabrications and properties*. CRC Press.

Pable, S., & Hasan, M. (2012). Ultra-low-power signaling challenges for subthreshold global interconnects. *INTEGRATION, The VLSI Journal*, *45*(2), 186–196.

Pable, S., Hasan, M., Abbasi, S. A., & Alamoud, A. R. M. (2013). Interconnect optimization to enhance the performance of subthreshold circuits. *Microelectronics Journal*, *44*(5), 454–461.

Poh, C., Mariatti, M., Noor, A. F. M., Sidek, O., Chuah, T., & Chow, S. (2016). Dielectric properties of surface treated multi-walled carbon nanotube/epoxy thin film composites. *Composites Part B: Engineering*, *85*, 50–58.

Pop, E., Mann, D., Wang, Q., Goodson, K., & Dai, H. (2006). Thermal conductance of an individual single-wall carbon nanotube above room temperature. *Nano Letters*, *6*(1), 96–100.

Pu, S.-N., Yin, W.-Y., Mao, J.-F., & Liu, Q. H. (2009). Crosstalk prediction of single-and double-walled carbon-nanotube (SWCNT/DWCNT) bundle interconnects. *IEEE Transactions on Electron Devices*, *56*(4), 560–568.

Qi-Jun, L., Zhang-Ming, Z., Yin-Tang, Y., & Rui-Xue, D. (2015). Influence of temperature on the conductivity of multi-walled carbon nanotube interconnects. *Chinese Physics Letters*, *32*(4), 047305.

Quick, R., Child, C., & Lanphear, B. (1895). Thermal conductivity of copper. *Physical Review (Series I)*, *2*(6), 412.

Rai, M. K., Arora, S., & Kaushik, B. (2017). Temperature-dependent modeling and performance analysis of coupled MLGNR interconnects. *International Journal of Circuit Theory and Applications*.

Rai, M. K., Chatterjee, A. K., Sarkar, S., & Kaushik, B. (2016). Performance analysis of multilayer graphene nanoribbon (MLGNR) interconnects. *Journal of Computational Electronics*, *15*(2), 358–366.

Raychowdhury, A., & Roy, K. (2006). Modeling of metallic carbon-nanotube interconnects for circuit simulations and a comparison with cu interconnects for scaled technologies. *IEEE Transactions on Computer-Aided Design of Integrated Circuits and Systems*, *25*(1), 58–65.

Redel, E., Wang, Z., Walheim, S., Liu, J., Gliemann, H., & Wöll, C. (2013). On the dielectric and optical properties of surface-anchored metal-organic frameworks: A study on epitaxially grown thin films. *Applied Physics Letters*, *103*(9), 091903.

Rochefort, A., Di Ventra, M., & Avouris, P. (2001). Switching behavior of semiconducting carbon nanotubes under an external electric field. *Applied Physics Letters*, *78*(17), 2521–2523.

Rossi, D., Cazeaux, J. M., Metra, C., & Lombardi, F. (2007). Modeling crosstalk effects in CNT bus architectures. *IEEE Transactions on Nanotechnology*, *6*(2), 133–145.
Rubio, A., Corkill, J. L., & Cohen, M. L. (1994). Theory of graphitic boron nitride nanotubes. *Physical Review B*, *49*, 5081.
Sanvito, S., Kwon, Y.-K., Tománek, D., & Lambert, C. J. (2000). Fractional quantum conductance in carbon nanotubes. *Physical Review Letters*, *84*(9), 1974.
Sapatnekar, S. (2004). *Timing*. Springer Science & Business Media.
Sarto, M. S., & Tamburrano, A. (2010). Single-conductor transmission-line model of multiwall carbon nanotubes. *IEEE Transactions on Nanotechnology*, *9*(1), 82–92.
Sathyakam, P. U., Choudhary, D., & Mallick, P. (2016). Mixed CNT bundle interconnects for subthreshold circuits. In *2016 International Conference on Information Communication and Embedded Systems (ICICES)* (pp. 1–4). IEEE.
Sathyakam, P. U., Karthikeyan, A., & Mallick, P. (2013). Role of semiconducting carbon nanotubes in crosstalk reduction of CNT interconnects. *IEEE Transactions on Nanotechnology*, *12*(5), 662–664.
Sathyakam, P. U., Karthikeyan, A., Rajesh, J. K., & Mallick, P. (2014). Reduction of crosstalk in mixed CNT bundle interconnects for high frequency 3D ICs and SoCs. In *2014 International Conference on Advances in Electrical Engineering (ICAEE)* (pp. 1–3). IEEE.
Sathyakam, P. U., & Mallick, P. (2011). Transient analysis of mixed carbon nanotube bundle interconnects. *Electronics Letters*, *47*(20), 1134–1136.
Sathyakam, P. U., & Mallick, P. (2012). Inter-CNT capacitance in mixed CNT bundle interconnects for VLSI circuits. *International Journal of Electronics*, *99*(10), 1439–1447.
Sathyakam, P. U., & Mallick, P. (2019). Triangular carbon nanotube bundle interconnects for subthreshold VLSI circuits. *Journal of Electronic Materials*, *48*(10), 6372–6381.
Sathyakam, P. U., & Mallick, P. S. (2017). Carbon nanotube interconnects with air-gaps: Effect on thermal stability, delay and area. *Journal of Nano Research*, *48*, 29–37. Trans Tech Publ.
Sathyakam, P. U., & Mallick, P. S. (2018). Reducing crosstalk induced delay and peak noise in carbon nanotube interconnects. *Current Nanoscience*, *14*(1), 76–80.
Sathyakam, P. U., Mallick, P. S., & Saxena, A. A. (2019). High-speed sub-threshold operation of carbon nanotube interconnects. *IET Circuits, Devices & Systems*, *13*(4), 526–533.
Sathyakam, P. U., Mallick, P., & Singh, P. (2020). Geometry-based crosstalk reduction in CNT interconnects. *Journal of Circuits, Systems and Computers*, *29*(06), 2050094.
Sato, S., Nihei, M., Mimura, A., Kawabata, A., Kondo, D., Shioya, H., et al. (2006). Novel approach to fabricating carbon nanotube via interconnects using size-controlled catalyst nanoparticles. In *Interconnect Technology Conference, 2006 International* (pp. 230–232). IEEE.
Shamiryan, D., Abell, T., Iacopi, F., & Maex, K. (2004). Low-k dielectric materials. *Materials Today*, *7*(1), 34–39.
Shulaker, M., Wong, H.-S. P., & Mitra, S. (2016). Computing with carbon nanotubes. *IEEE Spectrum*, *53*(7), 26–52.
Sieradzki, A., Pawlus, S., Tripathy, S., Gagor, A., Ciupa, A., Maczka, M., et al. (2016). Dielectric relaxation behavior in antiferroelectric metal organic framework [(ch 3) 2 nh 2][fe iii fe ii (hcoo) 6] single crystals. *Physical Chemistry Chemical Physics*, *18*(12), 8462–8467.
Sierociuk, D., Dzieliński, A., Sarwas, G., Petras, I., Podlubny, I., & Skovranek, T. (2013). Modelling heat transfer in heterogeneous media using fractional calculus. *Philosophical Transactions of the Royal Society A*, *371*(1990), 20120146.
Singh, K., & Raj, B. (2015a). Performance and analysis of temperature dependent multi-walled carbon nanotubes as global interconnects at different technology nodes. *Journal of Computational Electronics*, *14*(2), 469–476.
Singh, K., & Raj, B. (2015b). Temperature-dependent modeling and performance evaluation of multi-walled CNT and single-walled CNT as global interconnects. *Journal of Electronic Materials*, *44*(12), 4825–4835.
Sinnott, S. B., & Andrews, R. (2001). Carbon nanotubes: Synthesis, properties, and applications. *Critical Reviews in Solid State and Materials Sciences*, *26*(3), 145–249.

Sivasankaran, K., & Mallick, P. (2014). Bias and geometry optimization of FinFET for RF stability performance. *Journal of Computational Electronics*, *13*(1), 250–256.
Srivastava, N., & Banerjee, K. (2005). Performance analysis of carbon nanotube interconnects for VLSI applications. In *Proceedings of the 2005 IEEE/ACM International Conference on Computer-Aided Design* (pp. 383–390). IEEE Computer Society.
Srivastava, N., Li, H., Kreupl, F., & Banerjee, K. (2009). On the applicability of single-walled carbon nanotubes as VLSI interconnects. *IEEE Transactions on Nanotechnology*, *8*(4), 542–559.
Stan, G., Ciobanu, C. V., Levin, I., Yoo, H. J., Myers, A., Singh, K., et al. (2015). Nanoscale buckling of ultrathin low-k dielectric lines during hard-mask patterning. *Nano Letters*, *15*(6), 3845–3850.
Stavila, V., Talin, A. A., & Allendorf, M. (2014). Mof-based electronic and opto-electronic devices. *Chemical Society Reviews*, *43*(16), 5994–6010.
Subash, S., & Chowdhury, M. H. (2009). Mixed carbon nanotube bundles for interconnect applications. *International Journal of Electronics*, *96*(6), 657–671.
Subash, S., Kolar, J., & Chowdhury, M. H. (2013). A new spatially rearranged bundle of mixed carbon nanotubes as VLSI interconnection. *IEEE Transactions on Nanotechnology*, *12*(1), 3–12.
Tang, M., & Mao, J. (2015). Modeling and fast simulation of multiwalled carbon nanotube interconnects. *IEEE Transactions on Electromagnetic Compatibility*, *57*(2), 232–240.
Terao, T., Zhi, C., Bando, Y., Mitome, M., Tang, C., & Golberg, D. (2010). Alignment of boron nitride nanotubes in polymeric composite films for thermal conductivity improvement. *The Journal of Physical Chemistry C*, *114*(10), 4340–4344.
Terauchi, M. (2006). Electronic structure analyses of bn network materials using high energy-resolution spectroscopy methods based on transmission electron microscopy. *Microscopy Research and Technique*, *69*(7), 531–537.
Theis, T. N., & Wong, H.-S. P. (2017). The end of Moore's law: A new beginning for information technology. *Computing in Science & Engineering*, *19*(2), 41–50.
Torrens, F. (2004). Effect of type, size and deformation on the polarizability of carbon nanotubes from atomic increments. *Nanotechnology*, *15*(4), S259.
Tsai, M.-H., Tseng, I., Liao, Y.-F., Chiang, J.-C., et al. (2013). Transparent polyimide nanocomposites with improved moisture barrier using graphene. *Polymer International*, *62*(9), 1302–1309.
Tseng, I., Chang, J.-C., Huang, S.-L., Tsai, M.-H., et al. (2013). Enhanced thermal conductivity and dimensional stability of flexible polyimide nanocomposite film by addition of functionalized graphene oxide. *Polymer International*, *62*(5), 827–835.
Turkane, S. M., & Kureshi, A. (2017). Emerging interconnects: A state-of-the-art review and emerging solutions. *International Journal of Electronics*, *104*(7), 1107–1119.
Usman, M., Mendiratta, S., & Lu, K.-L. (2015). Metal-organic frameworks: New interlayer dielectric materials. *ChemElectroChem*, *2*(6), 786–788.
Vanstreels, K., Wu, C., & Baklanov, M. (2015). Mechanical stability of porous low-k dielectrics. *ECS Journal of Solid State Science and Technology*, *4*(1), N3058–N3064.
Verdonck, P., Wang, C., Le, Q. T., Souriau, L., Vanstreels, K., Krishtab, M., et al. (2014). Advanced pecvd sicoh low-k films with low dielectric constant and/or high youngs modulus. *Microelectronic Engineering*, *120*, 225–229.
Verma, R., Bhattacharya, S., & Mahapatra, S. (2011). Analytical solution of joule-heating equation for metallic single-walled carbon nanotube interconnects. *IEEE Transactions on Electron Devices*, *58*(11), 3991–3996.
Wang, H., Yuan, L., Liang, G., & Gu, A. (2016). Tough and thermally resistant cyanate ester resin with significantly reduced curing temperature and low dielectric loss based on developing an efficient graphene oxide/mn ion metal-organic framework hybrid. *RSC Advances*, *6*(4), 3290–3300.
Wang, J.-Y., Yang, S.-Y., Huang, Y.-L., Tien, H.-W., Chin, W.-K., & Ma, C.-C. M. (2011). Preparation and properties of graphene oxide/polyimide composite films with low dielectric constant and ultrahigh strength via in situ polymerization. *Journal of Materials Chemistry*, *21*(35), 13569–13575.

Wejinya, U. C., Chalamalasetty, S. N. S., Dong, Z., Arumugam, P. U., & Meyyappan, M. (2013). Carbon nanofiber nanoelectrode array: Effect of process conditions on reliability. *IEEE Transactions on Nanotechnology*, *12*(1), 101–107.

Wu, C., Li, Y., Baklanov, M. R., & Croes, K. (2015). Electrical reliability challenges of advanced low-k dielectrics. *ECS Journal of Solid State Science and Technology*, *4*(1), N3065–N3070.

Xu, S., Yang, Y., Pei, D., Zhao, X., Wang, Y., Sun, W., et al. (2010). A waveguide-like effect observed in multiwalled carbon nanotube bundles. *Advanced Functional Materials*, *20*(14), 2263–2268.

Yamada, T., Saito, T., Fabris, D., & Yang, C. Y. (2009). Electrothermal analysis of breakdown in carbon nanofiber interconnects. *IEEE Electron Device Letters*, *30*(5), 469–471.

Yang, M., Zhao, H., He, D., & Bai, J. (2016). Largely enhanced dielectric properties of carbon nanotubes/polyvinylidene fluoride binary nanocomposites by loading a few boron nitride nanosheets. *Applied Physics Letters*, *109*(7), 072906.

Yap, Y. K. (2009). *BCN nanotubes and related nanostructures* (Vol. 6). Springer Science & Business Media.

Yoo, B. M., Shin, H. J., Yoon, H. W., & Park, H. B. (2014). Graphene and graphene oxide and their uses in barrier polymers. *Journal of Applied Polymer Science*, *131*(1).

Zagorodniy, K., Seifert, G., & Hermann, H. (2010). Metal-organic frameworks as promising candidates for future ultralow-k dielectrics. *Applied Physics Letters*, *97*(25), 251905.

Zahedmanesh, H., Gonzalez, M., Ciofi, I., Croes, K., Boemmels, J., & Tőkei, Z. (2015). Numerical analysis of airgap stability under process-induced thermo-mechanical loads. In *Interconnect Technology Conference and 2015 IEEE Materials for Advanced Metallization Conference (IITC/MAM), 2015 IEEE International* (pp. 47–50). IEEE.

Zhang, K., Tian, B., Zhu, X., Wang, F., & Wei, J. (2012). Crosstalk analysis of carbon nanotube bundle interconnects. *Nanoscale Research Letters*, *7*(1), 138.

Zhao, X., Liu, Y., Inoue, S., Suzuki, T., Jones, R., & Ando, Y. (2004). Smallest carbon nanotube is 3 å in diameter. *Physical Review Letters*, *92*(12), 125502.

Zhou, Y., Johnson, J. L., Ural, A., & Xie, H. (2012). Localized growth of carbon nanotubes on CMOS substrate at room temperature using maskless post-CMOS processing. *IEEE Transactions on Nanotechnology*, *11*(1), 16–20.

Zhu, L., Xu, J., Xiu, Y., Sun, Y., Hess, D. W., & Wong, C. (2006). Growth and electrical characterization of high-aspect-ratio carbon nanotube arrays. *Carbon*, *44*(2), 253–258.

The manufacturer's authorised representative in the EU is Springer Nature Customer Service Centre GmbH, Europaplatz 3, 69115 Heidelberg, Germany. If you have any concerns regarding our products, please contact ProductSafety@springernature.com

Printed and bound by CPI Group (UK) Ltd, Croydon, CR0 4YY
30/09/2026
02225306-0007